MANAGING VIRTUAL TEAMS

SPIRO BUSINESS GUIDES
HUMAN RESOURCES AND TRAINING

Spiro Business Guides are designed to provide managers with practical, down-to-earth information, and they are written by leading authors in their respective fields. If you would like to receive a full listing of current and forthcoming titles, please visit www.spiropress.com or email spiropress@capita-ld.co.uk or call us on +44 (0)870 400 1000.

MANAGING VIRTUAL TEAMS

JOE WILLMORE

First published in 2003 by
Spiro Press
17–19 Rochester Row
London
SW1P 1LA
Telephone: +44 (0)870 400 1000

© Joe Willmore 2003

ISBN 1 904298 95 8

Reprint 2004
Ref 6501.JC.8.2004

British Library Cataloguing-in-Publication Data.
A catalogue record for this book is available from the British Library.

All rights reserved. No part of this publication may be reproduced, stored in a retrieval system or transmitted, in any form or by any means, electronic, mechanical, photocopying, recording and/or otherwise without the prior written permission of the publishers. This book may not be lent, resold, hired out or otherwise disposed of by way of trade in any form, binding or cover other than that in which it is published without the prior written consent of the publishers.

Disclaimer: This publication is intended to assist you in identifying issues which you should know about and about which you may need to seek specific advice. It is not intended to be an exhaustive statement of the law or a substitute for seeking specific advice.

Spiro Press USA
3 Front Street, Suite 331
PO Box 338
Rollinsford NH 03869
USA

Typeset by: Monolith – www.monolith.uk.com
Printed in Great Britain by: Biddles, UK
Cover design by: REAL451

Spiro Press is part of The Capita Group Plc

Contents

Acknowledgements	ix
About the author	xi

1	**Our changing world**	**1**
	Scenarios of virtual work	1
	Who this book is for	9
	What this book is and is not about	11
	Questions to consider	16
	Notes	17
2	**A horse is not a zebra**	**19**
	Understanding teams	21
	Comparing virtual teams and face-to-face teams	27
	Defining a virtual team	29
	Types of virtual team	32
	Stages of virtual team development	34
	Questions to consider	40
	Notes	41

3	**Different strokes for different folks**	**43**
	Differences in dynamics	44
	It takes every kind of people	48
	Virtual interaction and temperament	53
	Dealing with anger	64
	Manipulating the dynamics	68
	Questions to consider	72
	Notes	73
4	**First things first – starting up a virtual team**	**75**
	Confronting expectations	76
	Expectations about interaction	80
	Language matters	87
	Structural and role issues	92
	Questions to consider	99
	Notes	100
5	**Hang together or hang separately**	**103**
	The value of cohesion	104
	Building cohesion	105
	Understanding trust	107
	Trust and virtual work	111
	Building trust in virtual environments	114
	Strategies and activities	117
	What would you do?	123
	Questions to consider	125
	Notes	125

6	**Managing – a brand new day!**	**127**
	Out of control	128
	Outcomes not behaviour	133
	Coordination not control	139
	Communication as work	141
	Autonomy not isolation	143
	New skills for new work	144
	Questions to consider	151
	Notes	151
7	**New work order**	**153**
	What is telework?	155
	How not to telework	161
	Factors to consider	165
	Logistics	172
	What to do?	178
	Questions to consider	180
	Notes	181
8	**Facilitating and training virtually**	**183**
	Understanding facilitation	183
	Virtual meetings	188
	Videoconferencing as a meeting medium	193
	Chatting with your fingers	198
	Training online	202
	Virtual learning principles	206
	What to do?	211

    Questions to consider   213
    Notes   213

Bibliography   **215**

Acknowledgements

Any book is a labour of love – it involves a lot of mental labour and you'd better love doing it or it will drive you crazy. I must thank my editor Dr Glyn Jones for his interest, understanding, advice and support. Appropriately enough for a book on managing virtual teams, Glyn and I (for all our communication) have never met face to face. There are many others I owe a debt of gratitude to for their support and help. This book couldn't have happened without my clients who provided information, experience and practical application as we worked together to improve their virtual teams. All of my friends on ASTD's National Advisors for Chapters served as guinea pigs, sounding boards and team members. I thank them for their patience and energy, in particular Russ Brock who was a kindred spirit as we worked through virtual interaction issues for the NAC. There are many consultants who served as tremendous resources for me – too many to name. Even some consultants I've criticized in this book have still helped to advance the understanding of virtual teams significantly. However, I would like to acknowledge two particular peers who were invaluable – Lisa Kimball of GroupJazz and Trish Silber of Catalyst Consulting Team. Finally, the biggest debt of all is to my wife Cathy. She put up with my rants, ramblings and revisions while always encouraging me. I couldn't have done this without her.

About the author

Joe Willmore is President of the Willmore Consulting Group, a human performance improvement consulting firm located near Washington, DC in the United States. He recently served as a member of the Board of Directors for ASTD – the world's largest society of workplace learning and performance professionals. He has worked with a wide range of clients including: the World Bank, the Smithsonian Institution, the National Geographic Society, Lockheed-Martin, Intelsat and many other organizations around the world. When he isn't working, he kayaks and coaches football (holding the USSF National Youth License).

The author may be contacted at:

Willmore Consulting Group
5007 Mignonette Court
Annandale
Virginia
USA 22003-4050
E-mail: Willmore@juno.com

CHAPTER 1

Our changing world

Scenarios of virtual work

Nigel heads up a project team responsible for bringing a key product to market for his firm. Because of the need for quick development and production, his firm has formed a temporary alliance with a company from Finland that is usually a competitor. The deadlines are tight and there is fantastic time pressure. There is no time to create any permanent infrastructure or form a new organization – especially when the alliance will be just for this project anyway. Nigel needs to temporarily integrate the key players from each organization and get the product in on time and under cost to specifications. Without the usual organizational systems or structures and with such geographic separation and different cultures, Nigel will need to be adaptable and rely on technology to link his team members and get the work done.

Elena is responsible for pulling together a proposal on an important piece of work that her company is pursuing. Elena doesn't have the expertise to do it by herself but she doesn't have the time to bring all of the key personnel together face to face. This proposal has a tight deadline. Instead, she's going to have to pull everyone together virtually – and generate a proposal by a team of people at different locations. She'll solicit input from each of the subject-matter experts and then integrate their contributions into one document.

Isabelle is the head of graphic design for a consulting firm that has just expanded into the former Soviet Union. To better serve each of the field offices, headquarters has placed a member of the graphic design department in each office. Isabelle now finds herself managing a department with staff in six countries. She can't see any of her staff and yet must still manage their weekly performance. The geographic distances and the cost of travel are too great for her to visit each office more than once a quarter. Isabelle must create new expectations and performance measures to manage her geographically dispersed staff and use technology to interact with this virtual team.

Markus manages a small software game company. He needs good software programmers – without them his business would fail. So when one of his best programmers went on maternity leave and then decided to stay home to be with her child, Markus looked for ways to keep her as an employee. Two other key staff members have decided they no longer like the commute into the city. Unless they can work out an arrangement with Markus, they're going to quit and become consultants. And another programmer – Lev – is still recovering from a climbing accident. He'd love to be working but can't manage a full day's schedule and really needs to stay close to home. Markus is going to try to keep them all as employees – by adopting a telecommuting programme where they work from home.

Fahim is part of a team of analysts for a financial investment firm. He and his team members must track financial market performance and make recommendations on when to buy or sell holdings. The market (including stock prices and international exchange rates) is very volatile. A delay of minutes can cost the firm millions of euros. The only way Fahim can stay informed is through a network linking him with the other analysts and tied into the market. When he's away from his computer, he relies on his wireless phone for voice and text messages. His phone allows conference calls and during the course of a typical day, he'll average over 100 quick phone conversations with team members about decisions to buy or sell holdings. Regardless of where he is, Fahim is linked to his team-mates.

Onandi runs an engineering firm with branch offices throughout the Commonwealth. Each branch office has a general manager with key operational authority and autonomy. Those hand-picked general managers are critical to the success of their branch office – daily operations require their presence. Onandi can't afford to have them leave their worksites on a regular basis to seek advice and input from them on strategic issues – yet their perspective is critical both for quality decisions and organizational buy-in. The only way Onandi can regularly consult with them without pulling them away from offices in Kingston, Toronto, London, Glasgow and Liverpool is to meet virtually. Every day the team of general managers convenes by phone or videoconference to share organizational issues, discuss strategy and resolve priorities.

Kris is often faced with sudden deadlines involving work that must be completed, yet her company has a strong culture that insists upon a sane worklife and a reasonable workload. Overtime is frowned upon, relying extensively on consultants is too expensive, and temps have too much of a learning curve to quickly pick up the work in time to meet deadlines. So Kris has learned to let the work 'follow the sun'. Projects that originate in the London office are passed to the New York City office, then to San Francisco, next Tokyo, then Moscow before proceeding to the quality control function in Berlin. Reports that would typically take 18–24 hours of intense work never involve anyone spending more than an eight-hour day because of the transfer from office to office. Kris manages projects with teams of people that are never working at the same time and place, and rarely see each other.

Dand is manager of the sales force for a company that sells agricultural products. He manages seven sales reps (who are out in the field), three assistant buyers and an administrative assistant. His sales staff call in for information on products and pricing options necessary to close sales. The sales reps can't sell without the support of the buyers and all sales require the approval of Dand. None of them have been in the same room at the same time this month yet they just set a company record for sales. They interact constantly with each other using text messaging on pagers, phones and downloads to PDAs (personal digital assistants).

Murali is the shift supervisor for the telecommunications/call center of a mail-order clothing manufacturer. Because they have buyers around the globe (although most of their customers are centered in English-speaking countries), their work is literally 24–7. Murali must share information on sales trends and operational issues with the other two shift supervisors (who each work different hours). Yet the three shift supervisors are never in the same room at the same time – because of work schedules. Technology allows this supervisory team (and each of their shifts) to work seamlessly and coordinate with each other (as well as with shipping and customer fulfilment). There may be three different shifts in the organizational chart, but to the customer they are all the same team.

Did you find yourself described in one of these scenarios? Although your own professional life may differ in some regards from one of the stories above, you probably have a lot in common with at least one of them. These stories describe an increasingly typical work situation for many professionals today. This is a world that is fast-moving with tremendous time pressure, increasing complexity, lots of change, global competition and organizations and networks that are spread out or difficult to bring together face to face. All of these situations describe virtual teams. You may have heard that phrase (or other similar ones like 'virtual organizations' or 'geographically dispersed teams') described as just one more fad we see in the world today. But virtual teams and virtual interaction and work are no fad or passing fancy. We live in a world in which virtual interaction, especially virtually teaming, is becoming the norm and not the exception.

This is true regardless of the organization size and industry sector. While some firms may have more of a head start in technology or compete in markets with forces that encourage virtual interaction, all sectors and all types of organizations are moving in this direction. If you don't currently do work virtually, you soon will regardless of your business or profession. What forces are propelling us to more virtual work and technology-supported interaction?

- *Speed.* All organizations have shorter deadlines and pressure to operate faster. When I need the answer now that means I don't have time to get on a plane and fly to your office to meet with you. Instead, we trade e-mail or I call you up to get the information I need. Time pressure and the need for speed is forcing professionals to utilize technology to communicate.

- *Complexity.* The more complex our work becomes, the greater need there is to get input from other people (or subject-matter experts). Thus the more we must rely on teams to cope with this greater complexity. But because we don't need to rely on those team members continuously, it makes little sense to create permanent teams or permanent organizations. So instead those teams are linked by technology – intranets and phones and overnight mail – instead of being in cubicles next to each other in the same building. Or we establish alliances and partnerships that come together for temporary needs and than drift apart when the work is complete.

- *Fluid structure.* We see mergers and acquisitions, the formation of new institutions, organizational restructuring, temporary alliances and quick partnerships for the sake of specific projects spring up all the time now. Who has time to acquire a building, finish it and bring all the staff in when you can link them all by phone lines? In highly competitive markets, virtual work is a way to quickly ramp up production for firms that need to add new capacity.

- *Global competition and flatter organizations.* The case for multinational businesses using virtual teams to compete is obvious, but what about smaller, local firms? Before, your company may have served only Manchester. Now you serve all of the UK with branch offices throughout. As organizations spread their reach throughout a country, region or even the world more work will be done virtually. As firms get flatter (with managers supervising more staff and personnel being located at client sites), it becomes impossible to have everyone you manage in your line of sight. Consequently, many managers or team members now find that most of their staff or peers are not people that work right next to them, but instead are either on the road frequently or located at other sites.

- *24–7–365 work.* Any business with a website or call centre that takes calls outside their immediate area has discovered how quickly the workday expands and the work week never ends. This is more than just having a lot of work. Instead, a call centre located in Dublin may have its busiest hours at night – when customers from other parts of the world (like the USA or Australia) call in for help. Regardless of workload, in a global business, customers may be most frequent when your typical workday ends. Such work environments call for shift work (something hospitals, telecommunications companies and police forces have practised for decades). And in today's shift work, coordination and information sharing is crucial – thus a heavy reliance on virtual interaction.

- *Diverse workforces.* In today's world, human capital is crucial. Having smart, capable people is a huge competitive edge (just as not having them is a huge competitive disadvantage). Yet Western birthrates are dropping, making the talent pool of potential workers smaller. That puts even greater pressure on competitive businesses to attract and keep talent. An important method of doing this is to be able to reach out to non-traditional workforces and also keep good performers who no longer 'fit the mould'. In the past, a worker who left for maternity leave or because of a disability wasn't able to contribute while they were out. Older workers were often faced with the choice of full-time work or retirement. Now, technology makes it possible to integrate workers from their homes. It makes it easier to utilize non-traditional talent (like the physically disabled, semi-retired, part-time or student worker). It also expands both the geographic area and talent pool that companies can potentially recruit from.

- *Technology.* Simply put, advances in technology are expanding the ways we can interact with each other. These advances also make it possible for teams to work more effectively – if the technology is harnessed intelligently.

Who this book is for

While this book should appeal to many different kinds of people, I've written it with several specific audiences in mind.

- *Executives.* Most organizations accidentally back into virtual work settings. Through a series of small and unconscious decisions the organization inadvertently evolves until one morning the executive management team awakes to discover that the organization has become one with a high dependence on virtual functions, yet a failure to plan intelligently for this eventuality. Unfortunately, in such situations, the executives tend to discover there is tremendous discontinuity within the organization – all the work is being done in virtual teams, but appraisals are all done individually or real opportunities to capture knowledge and leverage assets are lost. They may have a series of virtual teams functioning in an environment that stifles the strengths of a virtual setting or with people working at cross-purposes in situations they are not prepared for or trained to function well in. While much of the material in this book is focused on a very practical level (such as tips for facilitating conference calls), I've made sure that the strategic implications aren't neglected. Thus executives will gain a better sense of the issues at stake for firms using virtual teams or evolving into the model. At a minimum, executives will be better prepared to make decisions that support virtual work (rather than operate like the proverbial ostrich with its head in the sand).

- *Managers and supervisors.* Virtual work and virtual teams create tremendous turmoil for many managers because most of the assumptions they have learned to operate by, to control work and to dictate performance are no longer true. In short, being a good manager in a face-to-face, traditional setting is very different from being a good boss in a virtual one. I'll explore all of these issues and provide a range of tips and tools for managers confronted with this changing world.

- *Facilitators, team leaders and team-building consultants.* Those who moderate meetings and help build strong teams may assume that what works for face-to-face meetings and teams should transfer just as well to virtual settings (with some allowances for the difficulties of technology). Nothing could be further from the truth. Virtual teams and virtual meetings have significantly different dynamics that alter a number of important facilitation approaches and team development. I'll cover a wide range of practical tips and tools, and implications for people in these categories.

- *Virtual team members and teleworkers.* Operating virtually involves a range of new challenges for team members and workers. One can't assume that management will pick up on these challenges and initiate changes accordingly. You need to take responsibility to raise issues to that you're prepared for a virtual working environment and your organization is prepared to support you. If you belong to a virtual team or are engaged in some form of telework (even if it isn't in a team), I'll point out a number of important considerations for people in these situations – both the workers as well as the organization.

Obviously, if you're currently engaged in virtual teaming or telework then this book is for you. But if you aren't currently doing this kind of work, this book is still very appropriate. People who don't do much virtual work at present will need to prepare themselves for more of it – a greater reliance on virtual teaming and telework is inevitable. Telework and virtual organizations are a paradox for many professionals and their companies. While the assets of virtual work are touted (and we see more virtual performance), there is also great dissatisfaction with it as well in many arenas as organizations blunder into telework totally unprepared.[1] Furthermore, while you can't prevent the increasing reliance on virtual work, you can shape what that virtual work setting looks like. So for those of you who wonder just what virtual work environments mean and what it is like to be on a virtual team, this book will help you prepare for those eventualities and give you the information you need to shape this organizational transition intelligently.

What this book is and is not about

From the previous section, you should already have a good sense of who this book is targeted at. But it is important to be clear about what topics and issues this book will and won't address.

Obviously, this book by its title will look at virtual teams. A quick definition of a virtual team is a group of people working together on a common goal who are separated by time or space and use technology to interact. We'll talk in more detail later about this (what we mean by virtual team and the 'flavours' that virtual teams come in). But this is more than just a book about virtual teams. To look at virtual organizations, cyberwork and teleworkers, one must understand virtual teams because teams are the basic building blocks of almost all organizational structures and the most basic form of collaboration and information-sharing. In order to look at virtual teams, we'll also need to consider their implications for the entire organization – because issues like organizational culture, communication networks, infrastructure and support systems all play major roles in how successful any virtual team will be. The role of managers and supervisors will also be a big topic – because virtual work environments change how managers manage. To be more specific, the traditional model of command and control isn't effective in a virtual world. Managers who are used to evaluating staff on the basis of their behaviour tend to find it difficult in virtual settings – because so much of the behaviour now becomes invisible to the manager. Performance management approaches will need to change with the onset of more telework. So this book will consider how managers need to change their managerial approach to succeed in virtual settings.

This book will also look at telework and its variations (such as hoteling). I know that many teleworkers don't consider themselves part of a virtual team or may not even be part of a face-to-face team at work. But much of what drives the development of telework leads to the creation of virtual teams. Additionally, many of the organizational dynamics and issues that arise with virtual teams are also present for telework. Plus, teleworkers will typically need to interact with co-workers back at the office. Even though most of the team members may be face to face, if some members (even if only one worker) are consistently separated by time or distance, then the team has virtual issues to consider. Finally, many organizations that end up with virtual teams start out by having one or two telecommuters or teleworkers. As the numbers grow, these virtual workers are eventually organized into virtual teams. So this book will take a good look at telework issues for individuals and organizations.

This book will also examine basic meeting, facilitation and training issues. Any workspace that formalizes telework arrangements or places a great deal of reliance on virtual workers will eventually need to look at how they meet (because meetings are usually one of the first ways that virtual issues arise within an organization). Training is critical for service and brainpower industries (and increasingly, most organizational sectors must rely on brainpower for competitiveness) so virtual organizations need to understand how training changes in a virtual world.

The nature of this book is to emphasize application, not theory. As you move on to subsequent chapters, you'll find the content becomes more specific and more practical. Most chapters will contain a series of tips and tools for troubleshooting work problems or improving a particular aspect of virtual teams. I will offer a number of short case studies and stories to illustrate how others have successfully (or unsuccessfully) dealt with many of these challenges. While I will reference a number of resources, the reality of virtual work is that new research and useful contributions on this topic are occurring almost daily around the globe. Thus it is inevitable that any resource guide on such a topic will be incomplete and should be viewed just as a starting point for any additional information you seek on this topic. There will also be questions at the end of most chapters for you to assess where you are and consider important aspects of virtual work.

This book is not a focus on technology. Too many people believe that virtual teams are primarily a function of technology. This leads to a belief that the team succeeds or fails to the extent it has good technology. Nothing could be further from the truth. Virtual teams may involve technology but that technology may be relatively crude (such as facsimile machines or conference calls) and the team can still be a high performing one. Likewise, the literature (and corporate grapevine) is replete with stories of teams with magnificent technological resources who failed resoundingly. It's natural for people to emphasize the technology aspect of virtual work. For some people, the technology or the hardware involves the toys or new playthings. For others, the technology is what they loathe. And the technology tends to be the obvious distinction – teams communicating by PDAs like Palm Pilots or Blackberries or using groupware – in how the work gets done.

But to focus on the technology is to misunderstand virtual teams. The truly unique aspect of virtual teams is not that they utilize technology. It is that we have changed the dynamics and nature of what we consider to be a team. For instance, many traditional, face-to-face teams use technology (such as desktop or laptop computers) to get their work done. That doesn't make them a virtual team. They are just a team that uses technology to do some of their work. A team of workers on the loading dock may use technology to get their work done but their reliance on fork-lifts doesn't make them virtual.

We now have groups of people working together under tremendous time pressure who must get information or input from each other but don't have time to physically get together. Or groups of people coming together, working and then separating – never to work together again. Or groups convening as complete strangers and expecting top-notch performance from the beginning of the team's formation. Or teams that are expected to collaborate despite different work cultures and different locations. Technology is only an enabling tool for these teams. What is special about these teams is not their use of technology (although some virtual teams do use technology very innovatively) but the nature of the team itself and how it differs from more traditional teams (where team members usually worked within sight of each other and almost always were members of the same organization both before and after serving on the team). As Lisa Kimball of GroupJazz notes, 'Although the technology which supports these new teams gets most of the attention when we talk about virtual teams, it's really the changes in the nature of teams – not their use of technology – which creates new challenges for team managers and members.'[2]

Besides, technology is always changing. So a book that focused on the latest and greatest videoconferencing hardware or the newest groupware would be out of date before the publisher saw the first draft. I will discuss technology (especially in terms of how synchronous or asynchronous technologies change the dynamics of how we interact) and how particular technologies enhance or penalize virtual teams. I'll also offer some examples of various virtual technologies in action. However, if you're looking for a book to tell you what version of a particular software package you should buy, you will be sadly disappointed here.

Questions to consider

Now that you've had a quick introduction to this topic, it's time to do some reflection before you move on to the next chapter. Please take a few minutes to consider your answers to the following questions. You may find it helpful to jot down your responses somewhere.

1. What assumptions do you have about virtual teams? In what ways do you feel working in a virtual team would be similar or different from that of a more traditional face-to-face team? How well do you think you would adapt to working from home (rather than in an office with co-workers)?

2. What obstacles or issues do you think your current place of work would need to change or confront in order to expand the degree of virtual interaction and virtual teaming that currently takes place?

3. What do you think would be the personal challenges for you to work in an environment that is more virtual than the one you currently work in? What difficulties would you have in adjusting to a more virtual work situation?

Notes

1. The Boston College Center for Work and Family found in 2001 that even as the percentage of teleworkers and telecommuters grows in the West, managers remain reluctant and distrustful of telework structures, employers feel their careers will be hurt by participating in telecommuting programmes, and many workers and organizations still struggle to make such programmes fit within their company culture. While there is strong evidence of the benefits of telework, it is not a panacea and many organizations have struggled with virtual work issues. The transition to virtual work settings isn't as easy as many organizations had assumed.

2. Lisa Kimball of GroupJazz has been one of the leaders around boundaryless facilitation, virtual teams and online conferencing. This quote comes from her white paper on virtual teams, available at her website www.groupjazz.com.

CHAPTER 2

A horse is not a zebra

There is a wide variety of horses. Horse enthusiasts can go into tremendous detail about different breeds and types of horses (such as Quarter Horses, Arabians, Appaloosas and so on). It doesn't take much knowledge of horses to admit that each animal also has its own individual traits – responding to each rider differently, showing aptitude for a challenge or skill that a horse of the same breed will lack the capacity to master. The science (or art depending upon one's perspective) of horse breeding assumes that the combination of lineage, diet, exercise and training can produce different results in otherwise similar horses. Yet, despite these individual and breeding differences, we can basically agree what constitutes a horse.

Let us now examine the zebra. It would appear that a zebra has much in common with a horse, both in basic anatomy, shape, biology and care. In fact, horses and zebras come from the same genetic family. One can even mate horses and zebras to produce offspring.

But if we push much harder at this comparison, we discover significant differences between the two. Horses have 64 chromosomes while zebras have only 44 (which means that only a male zebra may mate a female horse – male horses cannot mate female zebras – and any resulting offspring is sterile). Male zebras are not capable of mating until they are four to six years old (while male horses can mate much earlier). The behaviour of each animal is significantly different. In short, despite some similarity in appearance and genetics, the horse and the zebra are radically different creatures.

It isn't wise to assume that what would work with a horse (in terms of training, breeding, diet and care) would be equally appropriate for a zebra. Yet this is exactly what too many executives, managers, team leaders, trainers and facilitators are guilty of doing with regard to virtual teams. In other words they assume that what works with a face-to-face, traditional workgroup (a horse if you will) must work equally as well or almost as well with a virtual team (our zebra). To operate with such an assumption is to walk down a path to disaster. Not only do virtual teams vary dramatically from face-to-face teams in their dynamics and behaviour, it is even a mistake to group all face-to-face teams together into one category (except to distinguish them from virtual teams). There is an important implication with this statement – applying experiences that have worked in traditional workplaces (and with face-to-face teams) to virtual organizations and virtual teams is likely to be destructive and at least wrong-headed. To understand why this is true, we first need to have a basic understanding of what a team is and what makes them function.

Understanding teams

There are many possible definitions of what we might mean by the term 'team'. For our purposes, an acceptable definition would be: a group of people (small enough that all team members can recall each other) working to achieve a common goal that requires collaboration. Let us examine a few key concepts to this definition. The size of the team is operational. That is, if the group has so little interaction they are incapable of recalling the characteristics of each other minutes after they part, we would not call them a team. Thus teams can vary in size – as long as the size isn't so large that it starts to depersonalize the members to the point that some become invisible or unmemorable to the membership. The team must be working towards a common goal or objective. If we don't share a goal (but we each have our own agendas we seek to achieve), then the dynamics and behaviour within the collection of people begins to change and we're no longer a team. Finally, there must be collaboration. A group of people waiting at a bus stop have a common goal (to catch the bus). But that goal requires no collaboration (other than not attempting to all board the bus simultaneously!). This is nothing more than a collection or group of individuals with a similar goal.

Thus we can distinguish workgroups from teams. A workgroup may be a collection of individuals who work in the same department (auditors or a call centre). They may all have the same job title, the same hours and pay and work next to each other. But if the work does not call for collaboration (i.e. each person can do their job independently of the others and there is no need to share information or coordinate work) then there is no team. This distinction between workgroup (which is what exists in most organizations) and team is an important one because teams are capable of generating synergy. This means they can achieve more working together than the same individuals could achieve working separately. Additionally, teams are capable of building substantial levels of cohesion – much greater than that of workgroups. This cohesion – or social bonding – means that cohesive teams are capable of withstanding tremendous pressures and tend to cope better with difficult times (such as organizational trauma or long hours). When the going gets tough, groups or teams without cohesion fall apart and cohesive teams stick together. Finally, true commitment or ownership of a task occurs only in teams – not workgroups. Committed workers demonstrate far more attention to detail and pride in their work. This commitment or buy-in is capable only with a common goal.

I am not arguing that team members cannot have individual goals. In fact, it is common in any team for individuals to have their own (often private) agendas. Those agendas might involve goals such as: how can I get a promotion, or what do I have to do to get a better computer, or how do I get revenge on Clive for what he said about me in the meeting? But if it is truly a team, then there is a goal that all team members believe is worth doing and are serious about accomplishing despite their individual agendas. Additionally, while the team members may work apart from each other frequently, the goal must require some degree of collaboration or interaction. This does not mean everyone has to be doing the same kind of work. It means that the individual work is interrelated and dependent upon the results of others to achieve success. Thus if we are assembling cars and I am responsible for installing windshields, I am dependent upon your ability to correctly install the windshield casket in order to do my job well. I cannot succeed independently of your performance – we must work together (even though we may appear to be working independently of each other). If you install the window gasket incorrectly, I won't be able to install the front windshield (or it will fall out the first time the car hits a bump in the road).

Much work can be done effectively by workgroups. But there are some tasks where teams provide a substantial value to the organization over workgroups. For instance, work that benefits from true commitment or ownership is one where teams are more effective than workgroups. Work that benefits from collaboration (such as group creativity and multidisciplinary tasks) benefits from teams rather than workgroups. Work that calls for consensus and strong agreement is more successful with teams than with individuals or workgroups. It is also important to note that calling a collection of people a team doesn't make them one (just as failing to label an office or department a team doesn't mean they aren't capable of evolving into a team despite the organizational label within their firm).

However, not all teams are the same. Just as individual horses from the same mare and stallion will differ in personality, so does each team take on its own personality, norms and characteristics. But other than this individuality (a function of time, place and the individuality of the circumstances and membership), we may certainly group horses by breeds – which have common characteristics (under which each individual horse then demonstrates what makes it special or different from its siblings). We group horses by breeds because there are appreciable differences in performance and behaviour between all Clydesdales and Lipizzaners – despite the individual traits of each horse. This is also true with teams – there are several types or categories of teams, each with its own characteristics. While there are many different ways to categorize teams, for our purposes we can group them into four different types: teams with managers, self-directed teams, matrix teams and project teams.

Teams with managers usually consist of evolved workgroups. In other words, there is a manager who oversees the team. Typically, the role of the manager may differ slightly from that of a manager of a workgroup. Because of the collaborative nature of teams, managers of teams usually play less of a 'control' function and tend to focus more on enabling the team – removing barriers within the organization so the team can be successful. The team manager is often the information conduit to the organization or the 'point of contact'. The manager is usually asked to be more of a 'sounding board' for the team and help the team deal with 'outside' issues, but be more 'hands-off' when it comes to team operational matters.

Such teams are often mature workgroups where members acquire so much expertise (and the trust and confidence in their manager) that they work together effectively with little direction. In these cases, the smart supervisor learns to manage from a distance and to seek consensus (because the performers are so good at what they do). The role of the manager may also evolve in such situations. Initially the manager may be very hands-on and directive. But as the group members mature and demonstrate outstanding performance and evolve into a team, the manager may start to focus more on 'external relations' and enabling the team by fending off interference from the rest of the organization. Many such teams aren't labelled teams by their organizations. Or, you might identify five or six identical units (in terms of roles, staffing, function and resources) but only two of these units have evolved into teams and the others still function as workgroups. Usually this is because many of the work unit members are just 'doing their job' and haven't bought into a larger, common goal. Or the workgroup members don't collaborate consistently with each other.

With *self-directed work teams* (sometimes referred to as self-managing teams or autonomous teams), the team either selects their own manager (and this role is usually a rotating one) or the team shares the managerial roles (such as budgeting and scheduling). In such a structure, there are greater demands on the team members but they also have more ownership of the decisions within the team. Self-directed work teams often have problems connecting with their organization – with the absence of an obvious manager (or with rotating team leadership), others in the organization don't know who to contact in the team or who the appropriate person is for particular messages. Thus a very common problem with self-directed work teams is that they 'wither on the vine' within the organization. Cut off from access to information, they fail because of the organization's lack of support for this new way of doing things. Self-directed work teams are typically best with workforces that are very mature (in knowledge and expertise rather than mature in age) because self-directed work teams have more control over how their work is done, and work processes and operational issues that a manager might otherwise typically deal with.

Matrix teams are extremely common today because of the increasing work demands many professionals face – yet their organizations are unable or unwilling to add more staff. In a matrix team, team members serve multiple masters. Most or all of the team members belong to multiple teams (each with different areas of focus, different managers or team leaders, and different membership). Scheduling and coordination are initial problems. A greater issue for such teams is clarifying priorities and authority. Typically, high performers end up in great demand and may find themselves pulled in multiple directions because they are not only serving on multiple teams but have major responsibilities with each team. Thus matrix teams often have a feel of conflicting loyalties among the membership. Team leaders and managers, to be successful in a matrix situation, must be good at building loyalty and support because they are rarely in a position where they have complete authority over their team members. Matrix teams are often seen in consulting firms or professional service organizations (where most work is project driven, staff deal with 'billable hours' and there may be only one person in the organization with a particular expertise – but they are needed on five different teams for limited pieces of work).

Project teams involve work with a definable start and a clear finish or completion date. While much work these days involves projects, project team members typically come together for a limited time knowing that membership in the team will end. This has major implications for issues like trust, loyalty, organizational culture and supervision. The project team is also a mixture of contradictions. Typically, the project may have clear roles and responsibilities (while some projects may be chaotic, members are usually brought into the team because of a particular expertise required on that project). But the start-up and guaranteed end of the team means that culture and infrastructure may often be lacking. The team may have a temporary feel to it – because it is a temporary team! Simply put, project team members must be good at imposing structure on the task or project while comfortable with a lack of structure and permanence in the nature of the work. Members of such teams sometimes complain that this often feels like being self-employed – always wondering what will happen after the project is over and wondering when the next piece of work will come along.

Besides the nature of the team shaping its dynamics, each team also goes through development stages. In other words, a new team does not provide outstanding performance from the onset. There are typically initial stages where new teams stumble and seek to determine purpose and internal operating procedures. There are a wide range of models (such as Tuchman's scheme of forming–storming–norming– performing) to explain how teams develop. We won't focus on any of those models at this point because almost all of them are specific to face-to-face settings, not virtual ones. Suffice it to know that teams not only vary in their nature (the type of team they are) but they also go through specific sequences of growth or dysfunction.

Comparing virtual teams and face-to-face teams

At the end of the first chapter, I asked you to list assumptions you have about virtual teams and virtual work. One likely assumption you may have had (certainly one that many managers, executives and workers hold) is that face-to-face teams are much like virtual teams, especially in terms of their internal dynamics. Nothing could be further from the truth. There is a wide range of research that is beginning to emerge from a wide variety of fields that shows that team dynamics differ in virtual settings from those of face-to-face ones. We're beginning to discover that teams in virtual settings react to authority, idea generation and direction very differently than do face-to-face teams. Thus how managers direct their teams, how facilitators guide meetings, how trainers and OD (organizational development) consultants develop teams, and how team members and teleworkers interact with each other (as well as what works and what doesn't) is shaped by the internal dynamics determined by whether or not that team is virtual or face-to-face.

Let's look at just a few practical examples to illustrate this point (examples that we will cover in more detail in later chapters) that virtual teams and face-to-face teams have significantly different behaviours and dynamics. Take the issue of creativity within teams or groups. Most people would probably assume that a group or team interacting with each other face-to-face would generate more and better ideas than one in a virtual setting. Nothing could be further from the truth. Use the example of brainstorming. The largest number of recorded ideas for a face-to-face group trying to list as many possible creative uses for baking soda within three minutes is 83 (a group of Hewlett-Packard managers). A similar group (managers in the US government) doing the same exercise with the same time limits (but operating virtually) came up with 287 ideas.[1] Research by Rosalie Ocker (focusing on a mix of interaction modes) also had congruent conclusions – that virtual interaction seemed to enhance either the quantity or quality of ideas over those of purely face-to-face settings.[2] And it is a given that, when attempting to generate creative thinking within a group or meeting, it is important to withhold criticism. Yet virtual teams appear to respond well to criticism – in some instances creative productivity improves (contrary to face-to-face teams experiencing criticism of their ideas). In short, looking only at creativity, virtual and face-to-face dynamics appear to be quite different. We respond to the same behaviour or stimuli differently depending upon the circumstances (virtual or face-to-face) that we're in, and achieve different results.

But virtual teams and face-to-face teams differ in ways other than just creative thinking. How authority is perceived differs in virtual settings. Behaviour that is perceived as overbearing or dominating in a face-to-face setting is perceived as reassuring in a virtual one. How trust is developed and maintained appears to be radically different between virtual and face-to-face settings. How people form impressions of team-mates and methods of building the team are different in virtual teams. All of these issues will be examined in far more detail in subsequent chapters. But I mention these issues now to illustrate how different these creatures (a virtual team and a face-to-face team) are from each other – much like a zebra and a horse. I will leave it to you to decide which is the zebra and which the horse.

Defining a virtual team

Earlier, we looked at a basic definition of what constitutes a team (as opposed to a workgroup or a collection of people). How would we define a virtual team? For some people, they would argue that any team that uses technology is a virtual team. But this definition is too broad for our purposes – because practically every team uses some form of technology today to do their work (from computers to telephones to machines to electric screwdrivers or digital thermometers). For our purposes we can define a virtual team as any team in which the members are separated by time or distance or in which interaction is mediated by technology. Let's examine the implications of this definition. Of course, we assume the same characteristics of our earlier definition of team (common goal, collaboration, able to remember members). But the separation by time, distance or technology-mediated interaction is the new component that helps us distinguish the primary difference between a virtual team and a face-to-face team.

Another term or abbreviation that some organizations sometimes use for virtual team is GDT (or globally dispersed team). This term has some advantages in that it focuses not on the technology (the 'virtual' aspect), but the truly unique nature of the team (that it must behave like a team while being dispersed geographically). However, I tend not to use this term (GDT) because there are some aspects of virtual teaming we will be interested in that do not involve geographically dispersed team members. In other words, some virtual teams can involve team members actually being in the same room with each other at the same time (but interacting through technology). The baking soda brainstorming example mentioned earlier (which utilized group decision support software) is just one example.

Separation by time means team members must interact with a time lag. Thus if team members are in shifts and communicate with each other by posting notes on a cork bulletin board in their workspace (so each new shift sees the messages from the prior shift), they may be considered a virtual team. More likely, teams separated by time will communicate with asynchronous means such as e-mail or with documents posted on websites or intranets. But it is important to realize that the dynamics of a team that communicate with slips of paper because they work in shifts will be similar to the dynamics of a team that communicates by e-mail or electronic bulletin board because they aren't in the office at the same time. The unique nature of the team isn't the technology (the e-mail or the slips of paper), but that a group of people not in the same place or time who (in the past) would have been regarded as separate groups or teams must have the dynamics of a single team.

We also mentioned separated by distance in our definition. This concept is easily understood – the team members are dispersed geographically and don't work near each other. Thus even if they sometimes meet face-to-face, because much of their work and collaboration must occur over distances, this team is virtual. A team separated by such distance might rely on electronic chat or wireless phones or facsimile machines to interact. They could just as well use carrier pigeons to pass messages and they would still be a virtual team. Again, this is because the dynamics of the team are shaped by the necessity that they perform apart from each other, not that they use technology at times (or what kinds of technology they use).

Lastly, our definition used the phrase 'interaction mediated by technology'. This means that the team members communicate with each other but that most or all of the communication goes through technology of some sort. We can certainly picture teams in which members contact each other or send reports via the Internet or through videoconferencing. But can a team in the same room at the same time (in which everyone is capable of seeing each other face to face) be considered a virtual team? They can. There is one type of groupware referred to as GDSS (or group decision support software). Examples of GDSS include Ventana Groupsystems and CoVision's Council or Web Council. While this software can be used in distributed groups (i.e. groups separated by space) it is more typically used in a computer lab where all the team members are around a table with their keyboards and monitors. How can a team in which all participants are in the same room at the same time be considered virtual? Because the dynamics of a team communicating through technology are radically different than those of the same individuals discussing the same topic in the same room without the use of the computers and software. So, one important insight we must remember is that virtual teams don't have to be only teams using high-powered, glitzy technology to do their work. If the team members are geographically dispersed, unable to convene at the same time, or their interaction is mediated through technology of some sort, we can consider them to be a virtual team.

Types of virtual team

Earlier we covered different types of teams (self-directed workgroups, matrix teams and so on), pointing out how dynamics and behaviours differ according to the type of team. Not surprisingly, there are different types of virtual team. Of course there can be virtual teams that are self-directed or matrixed or established as project teams. And each of these versions of a virtual team would share some similar dynamics with a face-to-face team of the same type (such as a virtual matrix team and a face-to-face matrix team). But the element of virtuality adds some new wrinkles to the dynamics of each version of these virtual teams. For our purposes, their are three types of virtual team that are worthy of our attention because of how these types shape group dynamics and behaviour. They are:

- dispersed and electronic;

- mixed modes;

- insiders and outsiders.

Each of these types of team has a unique set of dynamics. A dispersed and electronic virtual team is what most people think of when they think of a virtual team. These are typically workgroups or teams that do all or almost all of their work and interaction through technology. Frequently, such teams are ones in which team members never meet face to face and never know what the other team members look like or even sound like (if they use text-based means of communication like e-mail or threaded discussion). A mixed modes team is one that has a significant percentage of work and interaction that is face to face (and a significant percentage that is virtual). What complicates this type of team is that even though the team is dependent upon virtual interaction to succeed, the team members may still impose face-to-face expectations and traditions on their virtual interaction. Later in this book, when we look at facilitating meetings, we'll see some clear examples of how teams need to interact and facilitate differently in virtual settings than in face-to-face settings. Yet mixed modes teams are usually guilty of assuming that all settings are the same and what works in one will work in the other. This is much like a football team that assumes that the tactics and formations that work at home will work equally as well for away matches because they're using the same players and same coach.

Obviously, such an assumption fails to take into account the difference in behaviour and dynamics between playing at home and playing away. An insiders and outsiders team typically consists of a situation where some members may be located together geographically and are able to interact and work face to face while some team members (contractors, part-time workers or staff based in the field locations) are forced to interact via technology. Thus those using the technology to connect with the team (such as calling in by phone to participate in a meeting everyone else is attending in person) tend to perceive they are treated like outsiders and feel excluded from many things in the team.

Stages of virtual team development

Face-to-face teams go through development stages. Virtual teams do as well. Despite the best wishes of executives and team leaders, any team (virtual or face-to-face) cannot instantly be a high performing one. All teams must go through initial stages where the team members learn how to work together and how to be an effective team. Thus it is so for virtual teams and their members. Given the special nature of virtual teams, it is not surprising that the developmental process for virtual teams also looks somewhat different from that of face-to-face ones. This following model (adapted from the work of David Sibbet[3]) explains the steps that virtual teams go through as they move from their initial formation to high performance. You'll also note that this model provides examples of the kinds of questions team members are asking themselves at each stage as well as diagnostic states to determine if the team has met that stage or not.

Virtual team performance model

Stage 1   *Orientation* – why am I here? why me? do I want to be part of this?
Unresolved – fear, disorientation, distance.
Resolved – acceptance, membership, clarity, engagement in the team.

Stage 2   *Trust building* – who are you really and what is your true agenda? am I with 'losers' or performers? is there hope for us?
Unresolved – mistrust, belief the team will fail or the members are inadequate, hidden agenda, undiscussable topics among team members, low energy level on the task, members avoid taking on tasks.
Resolved – group trust, member initiative to solve technical problems, spontaneity, information flow, integrity in communications with each other.

Stage 3 *Goal clarification* – why are we here? do we have anyone who knows why we're here? will I have to do more than my share?

Unresolved – apathy, irrelevant competitiveness, withdrawal, doubt.

Resolved – calls for action, options emerge, tasks identified, priorities set, high comfort level by team members, acceptance of the team leadership.

Stage 4 *Deciding the way* – how will we do this? how will we work?

Unresolved – dependence, conflict, poor fit appears to be the case as individuals feel they are doing work they aren't suited for or it appears the team doesn't have the right resources, unresolved differences continue within the team, unstated expectations not resolved.

Resolved – roles become differentiated, expectations on process are discussed within the team, work begins and members start to see results, decisions begin to be made, team alignment on details is achieved.

Stage 5 *Community* – we are one!

Unresolved – non-alignment, missed deadlines, conflict that appears to be personality-driven is frequent within the team, cliques or subgroups are obvious and well-established, many hidden agenda and political games exist that compete for time and energy with the team goal, some individuals feel they are outsiders who don't fit in.

Resolved – interaction in the team becomes intuitive and 'easy' for all members, members look forward to group exchanges and online sessions, unique team culture emerges, creative results, process is disciplined and consistent, team members look out for one another.

Stage 6 *High performance* – wow!
Unresolved – overload, non-attunement, uncoordinated work among team members leading to silos, high frustration levels among team members.
Resolved – synergy in action, flow, everyone owns the task, tremendous excitement or fulfilment from progress, problems and delays are seen only as obstacles to be overcome – not sources of frustration.

Stage 7 *Renewal* – why continue?
Unresolved – boredom, burn-out, interest or priority lags, team members are drawn to other teams or projects, cliques and social subgroups begin to flourish, members begin to deviate from the team expectations/ norms.
Resolved – commitment to renew, team membership and accomplishments become self-sustaining, congruence of vision, self and team.

There are a couple of points about these stages of development for virtual teams. First, not all teams ever reach stage 6 (high performance). In fact, a number of teams flounder at stage 1 (just as many face-to-face teams flounder at their initial stage of development). Second, these stages of development are a sequence. In order for virtual teams to be able to reach stage 5 or stage 6, they must first pass through the earlier stages in order. Also, it is certainly possible for teams to 'fall back'. In other words, a team might reach a stage and then engage in some dysfunctional behaviour and regress to a lower, initial stage and have to rebuild. Finally, one of the ways that virtual teams and their development varies significantly from that of face-to-face teams (besides the fact that face-to-face teams have different stages of development) is that if a virtual team is to progress, it usually does so in the initial stages very quickly or not at all.

This last point has several implications. One insight is that virtual teams may develop trust earlier than face-to-face teams do. A second implication is that virtual teams are capable of making major progress very quickly in their development – while face-to-face teams typically take more time to progress through the initial development stages. The last implication (and one that will be covered in far more detail in Chapter 5) is that virtual teams tend to develop trust (stage 2) very quickly or not at all.[4] Face-to-face teams, on the other hand, may take a long time to develop trust, but may still develop it. Trust is also usually further along in the development stage for face-to-face teams. For managers then, this means that one can usually tell fairly quickly (using this model of development) if a virtual team is going to develop or not. And if it isn't, it usually makes sense to 'pull the plug' since virtual teams that don't reach the first several stages in the development cycle within a reasonable period of time usually won't reach that far ever. Meanwhile, with a face-to-face team, patience is more in order. A face-to-face team may spend much early time spinning its wheels and floundering (or forming according to the Tuchman model), but may still eventually recover and grow. Virtual teams are more likely to grow quickly – or not at all.

There are several tips to keep in mind about these stages of virtual team development. First, this is a useful checklist for managers who must evaluate whether or not to keep a team going. As a manager, you can ask yourself how a new team is doing at particular stages. Even if the team appears to be having troubles or challenges, if they are progressing through the stages of development, then you should continue to have patience. If a new virtual team appears to be floundering in the early stages and haven't made progress, it should be a tip to end the team and start again.

Besides decisions such as that, it also serves as a checklist to make sure teams are on track and can also be used to help diagnose potential problems. For instance, a manager could assess a new team to see what stage they are at and to see that they are progressing well. Supposing a team is stuck at stage 2 (trust building), the manager will know that they either need to intervene quickly and strongly or conclude that the team won't be capable of generating serious performance. If a team is having trouble reaching stage 3 (goal clarification), then this provides some good clues to the manager as to what kinds of help the team needs to get back on track.

Finally, this is also a useful tool for team members. As the team progresses through each stage, team members (or a facilitator or team leader) can raise questions such as: 'What do we need to do to get to the next stage? What are the obstacles to us getting there?' For instance, a team just starting up needs to advance past stage 1 (orientation) and stage 2 (trust). Therefore, team members or the team leader should be looking at activities to deal with individual orientation and then team trust. While everything is important, these two initial issues (orientation and trust) need to be dealt with effectively for the team to be able to move beyond them and focus on more task-specific issues (such as goals, work processes and the quality of the performance). Thus this virtual team performance model is also a tool for helping teams focus where they need to spend attention in order to grow.

Questions to consider

Now that you've had a chance to get a better understanding about the differences between different types of teams, there are several questions that will help you apply this information to your own work circumstances.

1. Think of a virtual team that you've seen or participated in. Was there a particular stage of development that the team stalled at (temporarily or for longer periods)? What stages seemed to be the most difficult to reach?

2. What kinds of virtual teams, in your experience, do you see most frequently: dispersed and electronic, mixed modes or insiders and outsiders? Which type of virtual team do you think is likely to face the greatest challenges in order to be successful?

3. Can you think of tactics or activities that a virtual team could use to achieve each stage in the development model? What would those tactics or activities be?

Notes

1. The total of 83 ideas for face-to-face brainstorming came from the work of consultant Charles 'Chic' Thompson. For more details, check out his book *What a Great Idea* (1992). The total of 287 ideas came out of a brainstorming session with civilian employees and managers who worked for the US Navy using group decision support software to generate this total. In each case, the groups were given a total of three minutes to brainstorm as many creative uses for baking soda as they could possibly generate.

2. Dr Rosalie Ocker, now at Temple University, found that distributed asynchronous groups produced more and better solutions than face-to-face counterparts (as reported in Ocker et al., 1996). Research from several studies in Canada in the US by Brent Gallupe (e.g. Brent Gallupe et al., 1992) also demonstrated that 'electronic brainstorming groups have consistently been more productive than traditional brainstorming groups that operate under the same structured rules.'

3. David Sibbet is credited with developing the concept of graphic facilitation as a meeting and planning tool as well as the model for development of face-to-face teams which this approach is based upon. More information about his work can be found at www.grove.com.

4. Dr Dorothy Leidner and Dr Sirkka Jarvenpaa found in their 1998 research on virtual teams that 'trust in virtual teams tends to be established – or not – right at the outset. The first interactions of the team members are crucial.'

CHAPTER 3

Different strokes for different folks

In the previous chapter we looked at types of virtual teams (such as dispersed and electronic, mixed modes, and insiders and outsiders). Besides those particular types, there are some situational factors that influence the dynamics of virtual teams. In other words, while we might have two dispersed and electronic teams in the same firm, each team might get very different results and feel very differently to people who experienced each of the teams. This is because of the individual nature of each type of team. This individuality is shaped by many factors. The make-up of the team members is one factor. However, we already know from experience that a collection of smart, gifted people won't automatically (or even in most cases) result in a smart, gifted, high-performing team. And there are just as many cases of high-performing teams that consist of individuals who aren't regarded as particularly extraordinary at what they do.

Teams take on their own personality. The individual member characteristics help to shape that personality but are by no means decisive in determining what the team will look and feel like (let alone how it will perform).

Differences in dynamics

The list of factors which shape the nature and feel of the team is a long one. Obviously, the organizational culture is a significant factor. The nature of the team's tasks (and how motivating and realistic the goals appear to be) are other major factors to consider. The climate the team must function in (how competitive, trusting, open, serious, petty, bureaucratic and diverse, to name only a few aspects) will influence the team to some extent. But so far we have only named variables that would apply equally as well with face-to-face teams, not elements that are unique to virtual settings.

Imagine a matrix based on two considerations – time and place. Time and place are the two factors that have the most unique influence on virtual teams and their dynamics. Time and place are the primary factors that determine what technology the members of a virtual team use (or should use) to interact and collaborate. Time and place also shape the nature of the team. Time and place are the radical departures from traditional team structures because in the past you couldn't be part of a team unless you could either work together or meet together. This is important to note because the real uniqueness of virtual teams is not the presence of technology, it is how the nature of the team is changed. In short, we ask people who share no common corporate culture to work as a team. Or we ask people who have no face-to-face contact to work seamlessly. Or we ask people in different time zones and locations to work as if they were in the same building. The uniqueness of virtual teams is the new nature of teams, not the technology. What is challenging about virtual teams is not the hardware but the changing social contract and expectations we place on the team and its members.

One particular model for understanding how these variables (time and space) interact to shape group modes of communication was developed by Lisa Kimball of GroupJazz. That matrix or model looks like Figure 3.1.

Same time/Same place (Synchronous communication)	**Different time/Same place** (Asynchronous communication)
Same time/Different place (Synchronous communication)	**Different time/Different place** (Asynchronous communication)

Figure 3.1 Interaction of time and space.

Workers interacting with each other at the same time do so using synchronous (or real-time) communication. When you speak with someone face-to-face or on the telephone, that constitutes synchronous communication. Interacting at a different time (in other words, a time lag or delay in reception or response is involved) refers to asynchronous communication. For instance, 'snail mail' (letters or documents sent through the regular post) is an example of asynchronous communication.

Let us examine some examples of communication methods or technologies that fit in the matrix shown in Figure 3.1.

Same place/Same time (co-located, synchronous communication):

- Groupware/group decision support software (where participants are in the same room at the same time but interact by networked computers).

- Text messaging by PDA (such as a Palm Pilot transmission) to people in the same room.

- Some paper-based technologies (such as affinity diagrams, the nominal group technique, single negotiating text).

Different place/Same time (distributed members, synchronous communication):

- Electronic chat (text conversation through the Internet).

- Videoconferencing.

- Audio conferencing and phone conversations.

- Pagers (one-way messaging).

- Text messaging (by Blackberry or wireless phone).

Same place/Different time (co-located members – usually doing shift work, asynchronous communication):

- Bulletin board (electronic or cork versions).

- Storyboard.

- Libraries or learning centres (especially ones that use computer-based training where individuals may show up at different times to use the same computer monitor).

- Posted memos or reports (where team members come to the document).

Different place/Different time (distributed members, asynchronous communication):

- E-mail.

- Threaded discussion.

- Facsimile transmissions.

- Voice mail.

- Snail mail/postage or other forms of paper document distribution.

As technology develops, it begins to confuse the distinctions between some forms of synchronous and asynchronous communication. For instance, videoconferencing used to have a lag in its transmission capacity so that any attempt to carry on a conversation with much back and forth interaction was extremely awkward – much like attempting radio communication from outer space (with each message having a delay of several seconds before it was received). Some users would have insisted that videoconferencing at that stage of development was more asynchronous than synchronous. Because of the refinement of communication technology, some workers are actually able to push asynchronous technologies to the point that they operate like synchronous technologies. For just one example, think of individuals who have carried on nearly real-time 'conversations' with each other via e-mail while they were in neighbouring cubicles or offices. The speed of transmission, the comfort level exchanging with email and the bandwidth of their system meant that they were able to send and receive messages so quickly that the interaction took on the dynamics of a synchronous exchange. Such displays are becoming increasingly common and are helping to blur the boundary between synchronous and asynchronous communication.

There is one other note to point out about the previous comments. This book focuses primarily on teams and therefore the languages and references are about teams. However, these concepts and points apply equally as well to virtual organizations, workgroups who must interact with technology or individuals who are teleworkers or on their own. Thus an editor who works from home, a teleworker in a call centre, a management consultant who must travel to visit various client sites – all are individuals who may not be part of a team or workgroup yet would still find most of this material applicable to their virtual work circumstances. Don't let the language and references to teams limit your ability to apply these insights to your situation (whether it involves teams or not).

It takes every kind of people

Robert Palmer once sang a tune by that title and the concept applies equally as well to organizations. People are individuals and are different in so many ways. Even remarkably homogeneous organizations have incredible differences among their staff just because of personality traits. These differences affect how we work and how we approach our work, not to mention how we prefer to operate in teams and workgroups. For instance, Edward DeBono with his Six Thinking Hats tool identified a variety of distinct preferences that people display in their problem-solving stance within meetings or teams. DeBono argues that these tendencies often sidetrack groups or meetings because members may fall back on the preferences they are accustomed to or feel most comfortable with, regardless of how appropriate it is for the task at hand.[1] Thus one may end up with a group of critics who delight in picking ideas apart when the group's task is to generate multiple ideas. Or a collection of diverse group members may find themselves at cross-purposes with each other. While some members attempt to generate ideas, others seek data (not opinions) while still others are focused on metagroup issues (how should we proceed?) with the result being that the group may be talking about the same general topic but has no focus on purpose and direction.

Many researchers have attempted to categorize people by their personalities or preferences. One of the more famous bodies of knowledge is based upon the work of Carl Jung and has led to a preference or temperament sorter that is widely used and is called the Myers-Briggs Type Indicator (or the MBTI). You may prefer other schools and their perspective on human personality types. The purpose of this section of the book is not to advocate a particular school's approach to the study of personality and temperament. Instead, the reference to the MBTI is only for the purposes of illustrating how individual personalities may play out within groups and teams. Therefore let us use Jung, the concepts to come out of his work and the MBTI references because they are so widespread. If you prefer another model, you can just as easily apply it and the insights on individuality and personality will be the same.

According to Jung's research on archetypes (and subsequent work by many others including Otto Kroeger[2]), some individuals reflect temperaments that lead us to refer to them in part as ENs. An EN in MBTI langague stands for Extroverted-iNtuitive.[3] Generally speaking, an EN is someone who when asked a question figures out the answer by talking. Thus ENs tend to be quick to answer. They may often sound as if they change their mind as they talk – that is because they do. ENs figure out their stance or position by talking about the question or issue. On the opposite temperament or preference side would be the IS. The IS stands for Introverted-Sensor. Generally speaking, an IS person is much more reflective in terms of temperament. Ask a question to an IS and they generally think about it and compose what they want to say before they say it. Thus we can see that ENs and ISs tend to approach questions (and how they answer them) in radically different ways.[4]

Now imagine a face-to-face meeting with a mixture of people who are ENs and ISs. As topics emerge, the ENs quickly jump in, express their opinions and the conversation moves on. For the ISs, they find that once they've decided what they want to say, the conversation has passed them by and everyone has moved onto another topic. This does not mean that ENs and ISs can't work together – only that personality differences can often be a source of frustration (and the differences between ENs and ISs are only one small example to illustrate this).

As a facilitator or manager of a workgroup or team, these behavioural characteristics would be important to know. Unless a manager created opportunities for the ISs in her group to participate, she would be denied their insight and the ISs would find meetings to be a very frustrating and devaluing experience. So having some sense of the various personality traits of the people in a team is useful information to have.

But imagine a meeting (or team problem-solving interaction) in which all communication is asynchronous. Instead of asking a question and getting immediate responses (which would feel most comfortable for the ENs of the world), asynchronous communication (with its built-in lags – such as with e-mail or threaded discussions online) is naturally more reflective. Even if someone dashes off a message or a reply quickly, because the response to their own message is lagged, that provides time to reflect and reconsider. If you were in a face-to-face conversation with a peer or manager and stopped talking (and asked them to stop) for several minutes to allow you time to 'reflect' you'd probably draw some strange stares from those you had been talking with. But with e-mail or other kinds of asynchronous communication, reflection and a little time to think before responding is an automatic part of the communication process.

The implications of this are obvious. If you managed a virtual group of workers who were primarily ISs, they would work better and feel more comfortable using asynchronous technology to interact in most instances. Or imagine if you had staff that tended to act rashly and often made decisions that they then reversed the next day – after they'd had an opportunity to think about it further. A technology that was more deliberate and encouraged reflection would help manage their limitations and weaknesses. Regardless of the personalities with a task, if it involves a task that benefits from reflection and some time to think, then asynchronous technology enhances those elements of the interaction and helps get a better result. Thus some workforces (because of their personality types) are more suited for reflective (and asynchronous) technologies as a form of interaction. They are either more comfortable with the asynchronous nature or the delay in interaction works to counteract one of their limitations (such as a rush to judgment).

We've used the Myers-Briggs/Jung concepts and terminology for illustration because they are relatively widespread. But one does not need the MBTI or to accept Jung to be to be able to utilize these insights. Any model that distinguishes between personality preferences or work styles or problem-solving temperaments (such as the INSIGHT Inventory or Michael Kirton's KAI creativity index or any of the Carlson DiSC instruments as well as many others) allows managers to identify traits and then pick and choose virtual technologies that best fit those traits (or compensate for weaknesses).

Nor is this an argument that asynchronous technology is a better way to interact. Rather, technologies that virtual teams use for work all have different dynamics. An Internet chat session (which is synchronous but uses text) has a different feel and set of dynamics than does a threaded discussion (which is asynchronous) or a videoconference (which is synchronous). Particular technologies emphasize or encourage certain behaviours and discourage others. Therefore, a wise manager, facilitator or team leader can use the technology that a team has to manipulate and encourage particular behaviours and discourage others.

Virtual interaction and temperament

So far, we've looked primarily at personality traits and temperament. But technology can also be a successful way of minimizing some age-old forms of discrimination. For instance, groups that interact by text (such as e-mail, chat or threaded discussion) tend to focus more on content and less on appearance.[5] I've consulted with a number of virtual teams who had no face-to-face contact until after their project was complete. And then, with members seeing each other for the first time, I'd hear team members react to one of the team-mates with 'I didn't know you were black' or 'Wow – from your comments you sounded so much more experienced – I'd never have guessed you were so young!' I've had numerous comments from team members I've facilitated in which individuals privately expressed how the lack of a visual presence somehow meant their comments were given more credibility (or that they perceived that their age, sex, race, religion or nationality was less of an issue for the team). In short, some technologies result in members focusing more on content and less on the source of the content. Or the technologies hide characteristics that we often use as the basis for discrimination or are reflective of individual or cultural bias. Even in instances where team members already know what the others look like, the absence of a visual reminder seems to reduce triggers and encourage the team to react to the message, not the messenger. Thus, virtual technology may be a means of enhancing the integration of diverse membership – by gaining credibility and acceptance for people who might initially be judged by their skin, age, occupation or gender rather than their ideas and competence.

There is one key qualifier to this point about virtual interaction as a means of coping with some forms of discrimination. There is also some preliminary evidence that poor command of language may enhance discrimination in a virtual environment.[6] Thus I might not discriminate against you because of your accent, but if you write incoherently and spell poorly you will still face some bias from those who read your e-mails. This is not meant to say that effective virtual teams have perfect spelling or all members are superb writers. Instead, this is to say that in a virtual environment, some cues are invisible or minimized. If I can't see you, I pay less attention to the colour of your skin (even if I know it's not the same colour as mine). But if you display chronically bad spelling or incorrect use of the English language, that discrimination may still occur – because of what you write and how you write it. The implications for these dynamics are powerful – virtual interaction may be an effective means for levelling the playing field for various minority groups, allowing organizations and teams to benefit from diversity while minimizing some of the social baggage that various groups have to deal with.

Given that caveat, virtual interaction can be an effective means of compensating for group or team problems or enhancing traits that are valuable for a particular task. Some examples of group, meeting or team behaviour (functional or dysfunctional) which can be at least partially offset or mediated by various forms of technological interaction are:

- talkers who dominate conversation through their desire to talk, their skill at speaking, their quickness in response or their inability to be concise;

- quiet participants who tend to reflect before speaking, who may be intimidated by someone in the group or who may not be as boisterous as other participants;

- members who face various kinds of discrimination (from the obvious examples such as racism to more discreet behaviours such as role stereotyping);

- members who need solitude or quiet in order to do their best thinking;

- members who work best in multiple conversations at the same time;

- groups or meetings that have a poor memory, that have trouble capturing action-items and that often lose key content;

- groups that tend to have conversations that break down into disorganized streams of consciousness with little order or structure;

- groups with a mix of individuals who have difficulty focusing on the same issue and same task at the same time – but seem to work better engaging in concurrent work.

As I mentioned previously, technological developments continue to push the boundaries of what is possible and what particular virtual tools are capable of. However, it is possible to generalize about the kinds of activities that work best for particular technologies as in Table 3.1.

Before reviewing the contents of Table 3.1, please keep several points in mind. First, the examples within the table are only generalizations, not absolute conclusions. Despite that a fax exchange is not regarded an effective means for teams to brainstorm ideas, there are certainly cases where groups have succeeded in operating this way. So there are exceptions to the rule with each of the technologies listed below. Also, how effective or ineffective a particular technology is will vary somewhat with the circumstances. Thus file posting on an intranet (where team members go online and edit a common document – and the only version of the document is posted on the internal site) is an excellent way to produce a report. It's a mediocre way to design a motorcycle. Likewise, a telephone call from a sibling may be an effective way to keep you up to date on the goings-on within your family. But as a team leader, placing a series of individual telephone calls to each member of staff is not a very effective way of sharing information – unless there is a strong need to share the information on a one-to-one basis.

Let me also clarify some terms and concepts in this table. In all instances, the examples in the table are referring to activities within a team (rather than information-sharing between only two people or individual idea generation). Thus a telephone call may be an effective way to share information between two individuals but is inefficient within a team. The term 'information-sharing' refers to the process of getting information to team members, answering any initial questions and making sure everyone understands the details. 'Discussion' refers to an exchange, debate or dialogue within the team about a task (such as 'What is the correct sequence?' or 'What do the rest of you think about this plan?'). The term 'idea generation' refers to the process of team creativity – both quantity and quality of suggestions produced by the group. 'Group decision-making' refers to how easy or difficult it is for the team to find common ground, reach closure, weigh options or generate a consensus. 'Group product generation' concerns the ability of the team to produce a product or result of some sort (though it is likely to be a paper product – such as a memo, report, design, action-plan or implementation sequence). Generally speaking, 'group production generation' tends to involve some degree of sequence or steps rather than just a decision. Another thing that distinguishes group product generation from group decision-making is that the former typically requires some degree of precision and detail. Last of all, 'dealing with conflict' refers not to differences of opinion within the team on various issues or disagreement about work. Instead, it focuses on dealing with interpersonal differences – what to do when Harry hates Sally. I could have easily looked at other elements of team tasks (such as capturing knowledge or enhancing collaboration) but I have tried to choose general topics that would apply to most work teams.[7]

Table 3.1 Technologies and work activities

Technology	Information-sharing	Discussion	Idea generation	Group decision-making	Group product generation	Dealing with conflict
Voicemail (asynchronous)	Somewhat effective (as a one-way communication tool)	Ineffective	Ineffective	Ineffective	Ineffective	Ineffective
Telephone (one call at a time – thus it's synchronous in pairs but asynchronous for the team)	Somewhat effective	Ineffective	Ineffective	Ineffective	Ineffective	Ineffective
Audio conference (multiple parties on the same call – synchronous)	Effective	Somewhat effective	Somewhat effective	Somewhat effective (at reaching agreement with simple issues, but not complex ones)	Ineffective	Ineffective

E-mail (asynchronous)	Effective	Somewhat effective (best when messages contain previous text as a reference)	Effective	Somewhat effective	Somewhat effective	Ineffective
Bulletin board (asynchronous)	Effective (again, as a one-way communications method)	Somewhat effective (can be used to gather feedback)	Somewhat effective	Ineffective	Ineffective	Ineffective
Chat (synchronous unless someone is reading a chat transcript)	Somewhat effective (the nature of chat tends to result in superficial exchanges or very disjointed detail)	Effective (best if it involves simpler topics)	Effective	Somewhat effective	Somewhat effective	Somewhat effective

Technology	Information-sharing	Discussion	Idea generation	Group decision-making	Group product generation	Dealing with conflict
Groupware (synchronous unless someone is reading a transcript or backfiles)	Very effective	Very effective	Very effective	Very effective	Effective	Ineffective
Video conference (synchronous)	Effective	Somewhat effective	Ineffective	Somewhat effective	Ineffective	Ineffective
Video conference with text and graphics (synchronous)	Very effective	Effective	Somewhat effective	Somewhat effective	Effective	Ineffective

Virtual learning environment with audio, video, text, graphics and file sharing (like Centra or Lotus Learning Space – both synchronous and asynchronous)	Very effective	Effective	Very effective	Very effective	Effective	Somewhat effective (primarily when chat is used to limit exchanges, and there is a moderator to keep everything civil)
Listserv (asynchronous)	Very effective	Effective	Somewhat effective	Somewhat effective	Somewhat effective	Ineffective
Threaded discussion (asynchronous)	Very effective	Very effective	Very effective	Somewhat effective	Effective	Ineffective

Technology	Information-sharing	Discussion	Idea generation	Group decision-making	Group product generation	Dealing with conflict
Intranet file access (file posted on a website or intranet for members to access and edit – the usual result is asynchronous but it is possible for synchronous interaction)	Effective (if the information comes from one source, not many)	Effective	Somewhat effective	Effective	Very Effective	Ineffective
Facsimile transmissions (asynchronous)	Somewhat effective (if using a broadcast fax)	Ineffective	Ineffective	Ineffective	Ineffective	Ineffective

| Pagers and one-way wireless (synchronous or asynchronous depending upon the technology and usage) | Somewhat effective (low detail but good at providing quick, short information) | Ineffective | Ineffective | Ineffective | Ineffective | Ineffective |
| PDA/wireless text exchanges (while it could be asynchronous, typically this means of exchange is used synchronously for short messages) | Somewhat effective | Somewhat effective (unlike pagers, a quick exchange is possible) | Ineffective | Effective (only in forcing quick decisions, testing consensus or summarizing; ineffective in anything beyond this) | Ineffective | Ineffective |

Dealing with anger

You'll notice that none of the virtual team technologies scored well as a device for managing interpersonal conflict. Why do you think that is? Remember, we're not talking about task disagreements but issues involving personalities and anger. Typically, such 'social' emotions involve trust. Humans need non-verbal behaviour to help gauge trust in very trying situations. If Ian is angry with Gwen, her non-verbal behaviour when she speaks with him helps Ian decide how much he can trust her and how honest or open she seems to be. None of these technologies does a very good job of dealing with non-verbal communication. Some forms (such as conference calls) reveal some aspects of non-verbal speak (how our voice sounds) but deprives us of other elements (our posture, eye contact, body movement and facial expression). Video, which appears to come closest to approximating face-to-face interaction, actually is a poor substitute in this particular regard. Even though we might be able to see and hear someone in a videoconferencing session, we still unconsciously wonder what we're missing. Thus it adds doubt to an already inflamed or volatile situation. In an angry exchange, videoconferencing does not come close to face-to-face expression.

Ideally, in an interaction that involves interpersonal anger and distrust, if face-to-face interaction is not a possible method for addressing it (because the anger must be dealt with immediately and the team is too far apart to convene in person), other technologies that appear more distant would be a better substitute. The famed 'hotline' connecting the President of the United States and the Premier of Russia is not, as pictured in so many movies, a telephone line. Crisis management experts concluded that the times the 'phone' was likely to be used would be ones of great stress and distrust. In such instances, a phone conversation or video exchange might create more doubt than it would resolve ('why wasn't he looking straight into the camera – was he being deceptive?'). So instead, the infamous 'hotline' is actually a teletype line with keyboards at both ends. Messages are exchanged in text to minimize the risk of people misinterpreting behaviour or sounds when viewed or heard over distance and by a different culture as well as translation errors. This is also a good example of the value of precision during times of crisis. In many cases, face-to-face interaction is contextually richer but much less precise (and more open to misunderstanding) than that of text exchange.

We will examine trust within virtual teams in more detail in a later chapter. But since the issue of anger and distrust is of concern at the moment, suffice it to say that there are tips for virtual teams to deal with distrust and anger and upset. Some of these antidotes are common to face-to-face situations (good facilitation, professionalism, control of emotions). But the suggestions that are specific to virtual environments would be these:

- Use the strengths of the particular technology to deal with the conflict. If the team is in a text-based medium (such as e-mail or chat), it is easy to capture comments, archive them and at least minimize misunderstanding caused by poor hearing or a failure to perceive what another was saying. To put it another way, use the technology for what it is good for and don't try to use it for what it can't do (or doesn't do well). This may occasionally be frustrating for team members who may want to press on and address a particular topic when it may be better to defer it for the time being (until it can be dealt with in a more appropriate medium). Asynchronous technologies are terrible for spontaneous exchanges but good for those involving reflection and where you want to encourage people to think about what they say or mean.

- Be especially aware how much subtlety in communication is lost within a virtual environment. Humour, metaphors, understatement, sarcasm and intent are very culture-specific. Thus they are easily misinterpreted or missed until the team members become very attuned to each other. Emoticons[8] – using the keyboard symbols to help clarify a language such as this smiley face :) or this wink ;) – are helpful, but shouldn't be regarded as a significant asset in an angry situation.

- All virtual technologies have some limits of some sort (conference calls capture only sound, chat usually has a message length, e-mail is asynchronous, videoconferencing usually allows only one site to speak at a time – thus censoring the other sites until the first one finishes speaking.) So you should place limits on the discussion about the anger or interpersonal dispute. For instance, if you must use chat to try and resolve some interpersonal differences, acknowledge it. Set ground rules before the chat exchange starts – that messages need to be short and of a maximum length, that reciprocity is in force (before I can send a second message, you are allowed to respond – otherwise the result is a series of chat lines that interrupt each other with no flow – literally ships passing in the night!).

- Use a facilitator or a third party. Even if you're just talking about a pair of people with a dispute they'd prefer to keep among themselves, the facilitator or third party will prove invaluable in a virtual argument. The facilitator serves much like the role of a referee on a football field – making sure everyone follows the rules and doesn't get carried away. The facilitator should also periodically stop the exchange to test how everyone is and if the players think this is useful or the exchange should be terminated or changed in someway. The facilitator can also impose some reflection on the process ('before either of you send any messages, I want you to review the messages you've received so far and see if there is anything that, with hindsight, you'd now change'). Plus, with many virtual technologies, the facilitator can serve as a gatekeeper who can literally censor any messages that get too inflammatory or terminate the exchange if it starts to become counterproductive or destructive.

The point both of this chapter as well as the book is not that virtual interaction is a panacea – it isn't. Nor am I arguing that virtual interaction is equivalent to face-to-face communication or is a reasonable substitute. Instead, virtual and face-to-face interaction have significantly different dynamics. Consequently, each mode (although, realistically, virtual communication includes a number of modes of interaction) has its relative strengths and weaknesses for human work, especially within teams. Virtual technology enhances communication and work in a number of key instances. The enhancement is such that in some situations, given a choice between working face to face and working virtually, some teams should choose the virtual environment. But given the stressors of interpersonal conflict, virtual interaction isn't an effective medium for dealing with anger. Thus when we talk about the strengths of virtual teaming, it is also imperative that we never forget its limitations as well.

Manipulating the dynamics

Let's see how these concepts about group member temperament and technological dynamics can be used to enhance team performance and fit task challenges.

> Stephen is the manager of a team tasked with developing a staffing plan for a series of new field offices his firm is going to be establishing soon. On his team, he has representatives from sales, marketing, operations, customer service and human resources. All of these people are busy – and belong to other teams as well. Already Stephen's team is behind schedule (and they had an ambitious deadline in the first place). Whenever he calls a face-to-face meeting, half of his team can't attend because of other work commitments and the difficulty of getting to the meeting site (the firm has staff scattered all over London). Consequently, his team delays making any decisions (because of the missing members) or decisions they do make get challenged by the people who can't attend. They're getting further and further behind schedule and still have no reasonable work product to show management. How can Stephen use virtual meetings effectively given his current team situation?

What answers did you come up with to this situation? What would you advise Stephen to do (besides update his résumé)?

If you guessed that there are a number of virtual strategies that Stephen can use effectively in this situation, you're right. Stephen obviously has difficulty getting all the right people in the right room at the same time. Since it doesn't apparently work to act with some of those people missing, Stephen needs to look at asynchronous technology (like e-mail, listservs, threaded discussion, an intranet with file sharing) as a way of moving the team forward.

Stephen faces at least two problems in this situation. First, he can't get everyone there at the same time, yet still needs their input. Asynchronous methods of interaction will allow people to check in at various times and still provide input or make reports. Thus Stephen could send out an e-mail message to the team that a first draft of stage one of the staffing plan is on the internal website or intranet and everyone has 24 hours to provide their input (via electronic post-it notes) to the plan. After 24 hours, Stephen can take the feedback, incorporate it into the plan and provide the second draft for approval. Using asynchronous technology, people can still participate despite the scheduling conflicts. Instead of trying to get everyone together for the same hour, Stephen only needs to make sure people are available at some point during a 24- or 48-hour period. And ironically, the work may move quicker (because the search for appropriate meeting times may mean that meetings get scheduled two or three weeks from now – while the use of asynchronous communication means the draft goes out right now and feedback is expected back in 48 hours!).

The second problem Stephen faces is that he was on a tight deadline in the first place and things have only got worse with the 'wheel-spinning' and ineffectiveness displayed so far. The inability of the team to meet and make decisions has put them even further behind. This calls for a parallel work structure. In other words, the best way for the team to cover as much work as possible (while still paying attention to detail and involving everyone) would be to break the team up into subgroups and assign subsets of the task to each group. The work in each subgroup would then occur concurrently or in parallel. Each subgroup could use conference calls, e-mail, electronic whiteboards or even chat to work on their piece of the project (depending upon the task and the size of the subgroup). Each subgroup would then post their work (even in unfinished stages) on the intranet or some site available to the team (a threaded discussion, bulletin board, file-sharing site) so team members not part of that subgroup could 'visit', see the work in progress, post comments if they felt the need, and feel like they were part of the process.

This parallel process becomes doable with the virtual technology because each subgroup will work on its task yet have access to the work of the others. Typically, parallel sequencing of work tends to break down when those who aren't or can't be part of a subgroup (because they're on another subgroup) refuse to let the work proceed unless they can participate too.

So far we haven't seen any sign that this group has trouble reaching agreement when everyone is actually at the meeting. However, if that turns out to be an issue – that once the subgroups start generating work the entire team has some buy-in issues – Stephen can use groupware or group decision support software (like Web Council or GroupMind Express) to generate agreement virtually. Groupware (especially when it has structured processes like the nominal group technique or an affinity diagram or multi-voting) is especially effective at generating agreement and finding common ground.

You'll notice that the possible answers in this case didn't involve just one particular technology. Stephen ended up (we hope) using e-mail, groupware, an intranet with file sharing – a range of technologies. There is an important lesson here for virtual work. Organizations tend to use the technology that is at hand or that they have acquired and force the work to fit the technology. Albert Einstein once said 'if your only tool is a hammer you see every problem as a nail.' This is a common problem with virtual interaction. I once consulted with one American client who had invested in an expensive videoconferencing set-up for their organization. They insisted that all meetings use this videoconferencing capability (whether it was appropriate or not). In some instances, conference calls, faxes or an exchange of e-mail would have been far more effective (and faster). But the organization had the videoconferencing hardware, was proud of it, and was determined to use it whether it was appropriate or not. Therefore, the lesson is: don't use a technology just because you have it. Use it because it's appropriate for the task. Thus a group with several different tasks at hand might choose to deal with some of them in a conference call and assign others to be covered by e-mail rather than attempt to cover them all in a conference call.

I am not arguing that to be effective in a virtual environment you must acquire lots of expensive technology. It's truly amazing how many virtual teams (and organizations) out there succeed with very 'low-tech' solutions. And the opposite is also true – many 'technology-rich' organizations and teams flounder. And obviously, necessity requires that teams use what they have available. But the lesson from this point is that you consider the range of tools you do have available, rather than forcing a fit with the tool you prefer or the one you've invested the most money in or the one you're using at the moment. Successful virtual teams don't succeed primarily because of their technology, they succeed in part because of how they choose and how they use their technology.

Questions to consider

Here are some issues for you to reflect on as a way of gaining additional insight from this chapter.

1. Think of the meetings or teams you participate in. How could technology be used to compensate for weaknesses within that team? How could technology support your team or meeting to be more effective?

2. What virtual technologies does your organization have? Which ones do you feel are used effectively? Which ones are used ineffectively?

3. If your organization were to acquire a virtual interaction tool it currently doesn't have (or doesn't have widespread access to), what should that tool be? How could your organization benefit from it?

Notes

1. See Edward DeBono's book *Six Thinking Hats* (1999) or his website at www.edwdebono.com.

2. See Otto Kroeger's work on Carl Jung, archetypes and the MBTI in *Type Talk at Work* (2002).

3. Subsequent work based on Jung's foundational research (by Myers and Briggs among others) led to the Myers-Briggs Type Indicator or MBTI, which looks at different preferences and uses letters to help categorize those preferences. Thus an 'E' is an extrovert who gets energy from being around people while an introvert or 'I' is someone who feels drained and 'people-out' if they spend too much time in groups. Thus even through the term 'Intuitive' starts with an 'I', since that letter is already taken for 'Introvert', Myers and Briggs (and the subsequent MBTI literature) use the letter 'N' for the term 'iNtuitive'.

4. It is important to note that the work around archetypes and temperament argues not that people are always this way, only that these are tendencies and preferences. An EN is certainly capable of reflecting before answering, but 'thinking out loud' tends to be more natural for most ENs.

5. See Sirkka Jarvenpaa and Dorothy Leidner in 'Communication and trust in global virtual teams' in the June 1998 issue of the *Journal of Computer Mediated Communication*.

6. See Jarvenpaa and Leidner (1998).

7. For a different yet very insightful perspective on the question of which technologies fit which group tasks, see Deborah Duarte and Nancy Tenant Snyder's book *Mastering Virtual Teams* (1999).

8. For a more detailed explanation and examples of emoticons there are numerous online websites. These include: *The Smileys and Acronyms Dictionary*, *The Smiley Dictionary* and *Ultimate Chat List Emoticons*. The best emoticon online reference I've found is by Tracy Marks at www.windweaver.com/index.htm.

CHAPTER 4

First things first – starting up a virtual team

There are a plethora of issues to consider for virtual teams and virtual organizations to succeed. We've covered some of the conceptual issues that are important to a basic understanding of why virtual teams don't behave the same way as face-to-face teams. All teams, if they are to be successful, require a strong foundation. That foundation consists of a number of different elements that go into creating a strong team – or allowing the team to grow and become a high-performing one. It should therefore come as no surprise that virtual teams also have start-up issues. Now it's time to look at the practical issues involved with setting up a virtual team and getting it started on the right path – the path to success.

Confronting expectations

The vast majority of managers and staff expect a virtual team to be pretty much like a face-to-face team except that they'll have to use technology more often and the virtual team won't be as good – a pale imitation of what they could accomplish face to face. Not only are those assumptions wrong (as we discovered earlier in this book) but they raise the wrong questions. Of course a virtual team isn't the same experience as a face-to-face one. And for that matter, a face-to-face team isn't the same experience as a virtual one either. Thus a vital element of any successful virtual team start-up is to confront the many expectations and assumptions that the managers and team members have about virtual work. Otherwise, participants unconsciously expect a virtual team to be like a face-to-face one, only a little less so (face-to-face 'lite') rather than like a completely different breed.

The first (and most obvious) expectation to confront has to do with the assumptions people make about being on a virtual team. In other words, it is important to communicate upfront – before the team begins work – that people should expect virtual work to be a different experience. While you may not be able to explicitly spell out all the differences, it is useful to give a number of examples as a means of pointing out some of the ways that their work experience will be different. The examples you do choose to use will vary with the nature of the work (and degree of team experience) of the team members. For instance, individuals who currently commute to work (but will be transitioning to a team made up of people working from home) need to look at their experience and assumptions around a work day, work hours and when work ends – because people working from home tend to discover that the division between work and personal life gets fuzzier. It is too easy with a home office to allow work to intrude.

It is also appropriate to acknowledge that while you know it will be different for them, this is also a partially uncharted path. Researchers are discovering new insights about virtual work all the time. Partially this is because much of the virtual work that has happened previously has not been through design (where a company or academicians deliberately set out to look at virtual teams or virtual organizations) but through necessity – people needed to collaborate and they didn't have the time to get on a plane and fly to each other's locations. Because this has happened by accident or necessity and not design, many key lessons haven't been captured or exist only within the internal culture and memory of particular organizations. Additionally, the different types of virtual teams mean that the dynamics for a mixed modes team vary from those of an insiders and outsiders team. And the technology keeps evolving. Lessons learned from one organization may not apply to another because the nature of the teams is different or the technology used isn't similar. So the research in this field (virtual work) lags tremendously behind the practice. Thus you may very well discover insights and capabilities from the virtual workplace that are not mentioned in this book or aren't discussed within the research in this field. So, point out examples to team members of how life will be different but acknowledge that there are some differences that all of you will discover together. Acknowledging the uncharted nature of this virtual path is important because when the inevitable surprises occur, you want team members to accept them as inevitable rather than to regard the surprises as an indication of how management is unprepared or the team members can't adapt.

There is a great deal of literature in the field on virtual teams (including comments by the esteemed Charles Handy[1]) that fails to adequately challenge existing expectations around virtual teams. For instance, recent research has now shown that virtual teams do not need an initial face-to-face meeting in order to build trust (and that virtual teams build trust as well and in some cases even quicker than face-to-face teams[2]). Yet a number of popular books and leading consultants have claimed that, absent any face-to-face contact, virtual teams will fail. Or they claim that absent a face-to-face meeting, virtual teams won't develop trust or any cohesion – they'll feel cold and impersonal and will fail to bond. We know now that these claims are false. What drives otherwise reasonable and informed professionals to make such claims about virtual teams is their failure to understand their expectations about virtual work. If one expects virtual teams to be like face-to-face teams only with less face-to-face contact, then it is natural to view virtual teams as inferior. That is because virtual teams don't feel like face-to-face teams. If one expects that the only way to develop trust is through repeated face-to-face contact and then also being able to watch someone's non-verbal behaviour to assess if they're truthful, then of course a virtual team would appear deficient. But virtual teams have different dynamics – they develop trust differently. Individuals behave differently in virtual settings. So much of the literature that tends to view virtual work as a poor substitute for face-to-face performance is guilty of expecting virtual teams to be like their face-to-face cousins. With such expectations, of course they will appear inferior (much like if we believed our zebra to be just another form of horse and then concluded it was a poor horse).

Team members must go into the virtual experience realizing that this will be different from their face-to-face work. If they fail to understand this, they will always regard their virtual team as a 'poor sister' that is less enjoyable, less effective and less desirable than face-to-face options. They will complain more, work less and generally put the virtual work (if given a choice) last on their priority list. They will resent virtual assignments and they will look for opportunities to 'opt out' of such work.

I am not arguing that you must be a cheerleader for virtual work environments. There are certainly pluses and minuses to any work situation (virtual or otherwise). Different people and different tasks fit some situations better than others – everything is not equal. But people must go into virtual work situations with open eyes rather than closed minds. It appears that too many workers and organizations are stumbling into virtual arrangements without the necessary forethought and then complaining that their experience was negative. Throwing technology at people and hoping for the best isn't sufficient when it comes to virtual teams. Until we recognize (and get the team members to recognize) that the change isn't that they'll be using technology but that the nature of the team has changed, then we'll continue to experience lots of angst and frustration by everyone concerned.

Expectations about interaction

Besides the beliefs about what work in a virtual team will be like, there are many operational and logistical issues that need to be tackled. Many of the operational and logistical concerns are ones that any new team or organization (virtual or face to face) would have to deal with. However, there are a number of expectations or team norms that are either unique to virtual settings or are of more importance to virtual teams than face-to-face ones. One large set of norms revolves around expectations about team interaction.

Take the issue of e-mail as just one example. E-mail occupies a curious place in our pantheon of text materials. E-mail is not anywhere near as intimate as a handwritten and personally signed letter. Yet it is more personal than a memo or company report. Why does this matter? If you were to write a letter to me in your own hand and I then proceeded to pass it around to my team-mates, you might feel as if I had violated some rule of privacy or trust. Yet if you had generated a report on your word processor and sent it to me, you would have no problem with me placing it on the bulletin board by the cafeteria for all to see (unless of course, it contained proprietary information or you requested beforehand that the audience remain limited). So if e-mail isn't quite a report and yet not a personal letter, how do we feel about people passing e-mail around (or forwarding it) to others we didn't intend to send it to? Workers tend to regard e-mail as much less formal than other company documents (such as reports or memos). This is usually obvious in the range of materials that shows up on e-mail distributed around the company as well as the various salutations, closings and punctuation that is viewed as acceptable within e-mail. There is no clear consensus among workers in most organizations how to treat e-mail.[3] Consequently e-mail is often a tremendous source of frustration for many people who feel that it is misused by co-workers.

Most team members will have very different expectations about what is appropriate with e-mail (often expectations that are unconscious – they are not aware they hold these positions until they're in conflict with others on the team). For instance, just how widely should e-mail be distributed? Who should be on the cc: (or copy) list? Should one always reply or acknowledge receipt of a message? How often should one check one's e-mail? In business, many organizations talk about returning phone calls within 24 hours (on workdays when travel is not involved). What is the expectation for how soon e-mail should be answered? And there is the question of reciprocity – when one person devotes substantial time and effort to an e-mail message, how does it feel when they receive a 'got it' or 'I disagree' in response with no depth or further explanation? What messages are not appropriate for company e-mail and when is it permissible to use the company-wide (or team-wide) distribution list? The point is not that teams need rules or policies for all of these points. To the contrary – part of the value that teams provide over traditional workgroups is that teams typically bring flexibility and adaptability to an organization. One does not need rules and policies for everything (or even most things). But team members need to put these issues on the table for discussion to see which ones are major sticking points within the team. To put it another way, the team needs to be clear about what it needs to create norms about and what those norms should be.

Another expectation for teams to consider is the issue of when is 'off the clock'? To put this another way, is there ever any 'downtime' or away from work time for team members? One of the blessings of technology is that it makes it easier to reach people any place and any time. One of the curses of technology is that it makes it easier to reach people any place and any time. Most professionals have already experienced this curse – pagers that go off with urgent requests after hours, wireless phones that ring in cinemas, e-mails that arrive on weekends, conference calls scheduled with the field office in Tokyo that fit their time zone but mean those in Berlin or London are supposed to be asleep when they participate in the call. It doesn't matter if the postal service doesn't deliver on weekends – something can always be sent by e-mail attachment (any day or hour). A number of churches have now begun to ask parishioners to turn off cell phones and pagers before the start of the service. Some workshops and seminars have adopted policies of confiscating all wireless phones at the door (and giving them back at the end of the event).

It is because of problems like this that virtual work gets a bad reputation for many workers – a perception that technology is just a means for management to get more work from employees and make their day even longer. It does not need to be this way. As we will examine in later chapters, it is possible for work to 'follow the sun' so that organizations get development of products faster without people putting in longer days. But separate from the efficiency issue, virtual organizations and teams need to confront the realization that the technology they use to work and interact requires some kinds of limits. Team members need to look at the issue of what constitutes 'off the clock', not just in terms of billable hours or official schedules, but also when people are reachable or can be expected to respond to requests. If teams don't look at this issue, invariably one team member will shoot off an e-mail (forgetting the time zone and date implications) and get angry (and attribute it to unprofessional behaviour or poor motivation) when the receiver doesn't respond for three days. Because the e-mail got to the receiver's computer after the close of business on Friday and wasn't seen until Monday morning, the receiver may have felt they were 'off the clock'. Because it is possible to access the corporate e-mail account from home, the sender may have felt that a Friday afternoon message wasn't 'too late'. In most face-to-face settings, we don't need to have teams discuss when 'off the clock' is or what it means. But in virtual settings (or in face-to-face settings with the introduction of virtual technology), this discussion needs to happen or everyone will discover they have no downtime – that work never ends.

As senders, team members often use technology to their convenience (sending a late-night e-mail from home when they can't sleep, placing a phone call from the taxi on the way to the airport, working at the weekend to avoid interruptions and then calling co-workers at their homes to find an answer to a question they need to complete their work project). Yet, while we may use technology when it's convenient for us as a sender, we often don't consider if it's convenient to the receiver, for example the person who places a call from the taxi headed for the airport and leaves a voice mail then turns their phone off when they get on the plane (which means the return call goes to their voice mail – and a vicious cycle of telephone tag is started). So another key issue that new teams need to look at has to do with expectations around reachability and when people are on or off the clock when it comes to technology. As simple as this issue sounds, it is truly amazing how much frustration, anger and misunderstanding results from unstated expectations around this issue.

The reverse of the 'off the clock' expectations involves when people are on the clock. As silly as this initially sounds, a major issue for some teleworkers and professionals who must interact virtually is their failure to make time for virtual events. Take the example of an online conference. Probably the biggest barrier to participation for online conferences among people who initially indicated they would participate is they simply fail to allocate time to sign on. This would be equivalent to someone promising to attend a face-to-face meeting in their office but then failing to put the meeting in their schedule or not leaving time to attend the event. At the end of the day, they're left bemoaning about how busy they found themselves to be. In short, people generally fail to treat virtual commitments like face-to-face commitments. Or the virtual commitment is not treated as a real commitment. A failure by people to understand their assumptions about these events means that virtual commitments get treated as if they require no time. Instead, people need to treat virtual commitments like face-to-face ones. This means, they need to schedule them and leave time for them. They may need to put a sign on their door that says 'in a meeting' or 'important conference call' in order to ensure that the virtual meeting isn't subjected to interruptions that wouldn't happen if they were out of their office attending a meeting in person.

It is important for teams to confront the expectations members have about a range of issues. What is particularly challenging about virtual work is that team members without much virtual experience (or teams where members have no experience with each other) will all have plenty of assumptions about how things should be done operationally, but they will not be aware that they have these assumptions. Those assumptions are tacit or unconscious rather than explicit. Thus Ian will expect that when you get his message, you'll acknowledge receipt. So when Pam doesn't (because her expectation is that with so much e-mail going back and forth, one doesn't reply to messages just to indicate you received it – thus creating even more e-mail), then Ian is likely to view this as unprofessional behaviour or poor motivation on the part of Pam. In short, a simple difference in operational expectations serves to generate perceptions of incompetence or poor motivation. Thus, it is important to make tacit or unstated expectations into explicit ones (that the team has at least acknowledged or confronted). Thus even if the team can't reach agreement on what to expect regarding e-mail protocol, at least members will realize there are differences of opinion on that issue and they won't regard deviance from their own standards as a betrayal or lack of professionalism.

Nor is it sufficient to simply ask 'what expectations do we have about working virtually?' Because most of these operational assumptions and expectations are tacit (or not obvious), that means members aren't conscious that they hold these expectations. If you ask team members what assumptions they're making about virtual work, they're not likely to be able to answer with much detail. It's much like asking the question 'What is it we don't know that we don't know?' Therefore it becomes important for the team leader or manager (or even a team member if the leadership fails to do this) to raise questions about specific issues:

- Are there any times it is unreasonable to expect a reply?

- What protocol will we have with e-mail (or any other means of interacting virtually)?

- What behaviours or uses have you seen with e-mail (or whatever the particular technology we're using) that you found especially irritating or frustrating?

- What expectations do we have around returning phone calls or updating our voice mail?

- What guidelines do we have (or we should create) around content and format for messages or meetings?

- How do we control the amount of traffic (so we don't restrict content yet don't drown in messages)?

- How do we indicate priority and urgency (so vital requests aren't lost in trivial questions)?

As trivial as these issues sound, they shape the experiences team members have of their team. They influence what perceptions the members form about their peers (and how professional, committed and knowledgeable they appear to be). Furthermore, when members experience a conflict with unstated expectations, they tend to attribute the differences not to misunderstanding or different assumptions but to unprofessional behaviour or poor motivation. In short, something simple is interpreted as a sign of something major. Failure to address these kinds of expectations means that teams typically blunder through their initial stages. And if you remember the stages of virtual team development that we discussed in Chapter 2, virtual teams that flounder early may never recover. Additionally, team norms (informal, unwritten rules or codes of conduct) tend to be developed inadvertently or through accident (rather than intentionally and with purpose) in most groups or teams or organizations. Not only are these norms mostly the result of accidents but they are shaped primarily on the basis of initial behaviour. If the team (out of ignorance or caution) demonstrates particular operational behaviour early, there is a risk that such behaviour becomes cemented as group or team norms – the way things get done within the team regardless of how counter-intuitive they may be. Thus, the team may build in inefficient or counterproductive behaviour.

Language matters

When I say that language matters, I'm not referring to basic literacy or the ability to master words or vocabulary – obviously those are basic elements of communication. But individuals who all speak English may be guilty of not speaking the same language. Professions have their own jargon which allows precision or saves time when dealing with peers but may be impenetrable to outsiders.

Many professionals (especially within the West) use metaphors as a means to enhance communication. Humour is often culturally specific (and culture in this context can refer to subgroups such as software engineers or government budget analysts or academic administrators). These points are important considerations because humour and metaphors often don't communicate well in face-to-face situations. But when attempting them over distance through technology, the possibility of misunderstanding is even greater. This is complicated by the nature of today's workforces – there is far more cultural diversity in our organizations and among our clients (not so much people from other heritages than our own, but a greater likelihood that we will serve on a team with people from engineering, finance, marketing, productions and field operations – all distinct organizational cultures of their own).

This concern about humour, metaphors and language is easily illustrated with an example from a client who shall remain nameless (or to paraphrase the television show Dragnet, the names have been withheld to protect the guilty). I once took part in a conference call involving a client headquartered in Toronto, Canada and with branch offices in Mexico City, London, New York City, Dallas and Paris. Despite the international nature of the firm, all the participants (principals in the firm) were fluent English speakers with an excellent command of the language. Furthermore, all had worked for many years in this firm and were well immersed in the firm's own culture and knew the argot of their work well. The call itself took place in English. As the managing partner from headquarters in Toronto proceeded through the call's agenda, he asked the principal in each office to provide an update on any big events coming up in the week. When it came time for the Mexico City office to share, the principal from that office mentioned that they had a big meeting scheduled with the firm's longest-standing client in Mexico to discuss an ongoing project. When asked by the managing partner how that meeting should go, the Mexico City principal responded with 'we expect it to be a superclassico'. Everyone else in the call evidently took the term as I did at the time – 'superclassico' must mean a combination of super and classic – a good thing one would think. We found out only a week later after the meeting had occurred that a 'superclassico' is a football term popular in Mexico (and which has now spread through most of Central and South America) that refers to a grudge match between two long-standing rivals – much like the Rangers–Celtic Old Firm match in Scotland. In short, the meeting was being described as anything but super and classic. In this instance, the meeting was a highly combative and antagonistic affair – not the sort of meeting one wants to have with a long-standing client over a major project. The problem here was not one of language – everyone in the meeting spoke English and a few of the principals also spoke some level of Spanish. The problem was the combination of the use of metaphor (the football grudge match concept – a term that would have been obvious to everyone in the Mexico City office) with the inability of anyone else to see the speaker's non-verbal behaviour, which might have provided a clue that perhaps a 'superclassico' was not such an ideal status for this meeting. Again, the point is not that technology prevents clear communication – there are plenty of examples of miscommunication in face-to-face interaction. The point is that virtual teams are more likely to have distinct and different cultures (which means metaphors and jargon are not shared) and team members can't rely on observation to detect if non-verbal behaviour is inconsistent with the verbal content – something that might provoke a question or two by an observant listener. Thus virtual teams communicate best when they leave metaphors and humour out of their messages. This is not because organizations must be politically correct or worry about offending anyone. But because so much humour and metaphorical speech is culturally specific, it is easily lost or confused with a diverse population or a team where face-to-face contact is missing.

Related to this issue of language is that of community. Amazingly enough, virtual groups with no face-to-face contact are capable of developing strong bonds and a tremendous sense of community. You have probably heard of stories involving very committed, strong, healthy marriages where spouses suddenly leave children and their mate to form a new life with someone they know only from the Internet – no face-to-face contact or interaction of any kind other than through e-mail or chat. Very strong bonds are capable of being formed through virtual interaction – bonds that are capable of outweighing those created face-to-face. We will explore how trust and cohesion develops in virtual teams in the next chapter. From a virtual context, community refers to a sense of belonging and of unity that exists despite a lack of face-to-face contact or tremendous diversity and heterogeneity among the participants. When community develops in a virtual group, individual members begin to take responsibility. Members take the initiative to deal with problems and they worry about other individuals who might be having trouble. They help each other out, they bond and an online community develops. The group is transformed from a collection of individuals to a collective – a perspective that each individual belongs to something greater than themselves. This sense of community is especially powerful for a team because individuals place even greater priority on team or group goal welfare over their own individual status. When an individual doesn't participate, others seek them out. When a sense of community is created, members look out for each other.

While we will look at community and trust later, a key point to remember with virtual team start-ups is that initial interaction within the team tends to become 'cemented' and develops into norms. If the team members initially interact very haltingly or awkwardly (because no one is clear about procedure or no ground rules exist to clarify expectations) then an unwritten rule emerges within the team. If initial meetings have no responses when the team leader calls for suggestions or ideas (because team members are still figuring out how to use the response key in the groupware software), then they start to perceive even after they become more competent with the software that online conversations are supposed to be slow and not very interactive. The hesitation to respond becomes a norm within the group. If members have trouble participating in the conference call (because they have trouble calling in without being cut off from the call) so initial participation is low, the norm that emerges is: we don't expect full participation in our calls – it's fine to give them a miss. So it is important to remember that early behaviour within the team helps create norms that cement that practice into an unwritten rule for the team. Thus it is important to be clear about what behaviour you do want within the team and work to establish it early on in the process. If community is vital to your virtual team, you'll need to lay the groundwork so that very early in the team's experience, they start to develop community and act in ways that are consistent with a sense of the whole (rather than a sense of exclusion).

Structural and role issues

Of course any organization or workgroup or team, be it virtual or not, requires some degree of structure and definition of roles. Therefore, it should be no secret that this is a major start-up issue for any virtual team. It's valuable to allow teams leeway in how they define roles or what structure they adopt. It's also realistic (and desirable) to let those roles and structures evolve as the team gets more experienced with the task and talent involved. But some degree of planning and definition is necessary at start-up.

One of the first structural components that is extremely useful for any virtual team is a team charter. A team charter makes sense for any work-related team – virtual or face-to-face. A team charter typically covers the team's mission or purpose, some hierarchical data (who the team's champion is, how team leadership is handled) and information about operational agreements and roles. The charter is usually a repository and formal commitment of team norms as well. This means that the charter often covers:

- how conflicts will be resolved and how decisions will be made;

- initial goals necessary to fulfil the team's mission;

- codes of conduct or expected behaviour or ground rules;

- individual or subgroup areas of responsibility.

In short, the team charter is both a commitment by team members to a number of key operational and strategic issues, and also a resource for both team members and those outside the team to refer to.

A second structural issue important for any team, but especially vital for virtual teams, has to do with the issue of team roles. One key role is that of POC or point-of-contact. When a team is rarely together or is usually out of sight visually, contact and information dissemination (especially from others outside of the team) becomes haphazard or non-existent. People just aren't sure who to contact on the team or who the appropriate person to send a message to is. Thus they tend to send it to everyone (contributing to information overload and e-mail clutter). Or they tend to send it to the wrong person (who does nothing with it because they don't know the significance of the message or assume the right person on the team also got the message). Or they don't send it at all (because it isn't clear to them there is someone on the team responsible for this issue or that the team is the correct repository for the information). Or a virtual team may be invisible to others in the organization, so they don't think to include the team on the distribution list for key material. Consequently, virtual teams need to establish a clear point-of-contact and provide that POC with some visibility. This is not necessarily the manager or team leader (who may be difficult to reach because of travel or may have little involvement in day-to-day operations within the team and thus not realize the significance of the message). The POC may not even be a person – it could consist of a button on a team website or a link to an e-mail box for people to direct messages to. The point is that the team needs to have an obvious and visible contact point for people to send team-related information or queries to. In the absence of this, the team will find themselves missing key information.

Another key function within a virtual team has to be that of knowledge manager or archivist or scribe. Typically, organizations do a terrible job with corporate memory or of capturing insights on the job. Think of all the meetings you've attended where – after the minutes are disseminated the week after the meeting (and you're lucky if it only takes a week to get the minutes back) – participants react with 'that isn't what I agreed to'. Likewise, you can probably recall instances where groups reconvene after some time apart and team members can't remember what the action items are or what agreements were made in the last meeting. In short, stuff happens, decisions get made and they get lost or forgotten. Organizations find themselves continually having to relearn important issues or skills. Related to this, with lower retention rates in most organizations, there is a greater need to provide training and information to new employees to bring them up to speed. Given almost continuous turnover, that issue (orienting new staff and preparing them to be productive) has become a significant challenge for almost all organizations.

What this means for virtual teams is that each team needs to be designed with the assumption that before key tasks are finished, the team is likely to lose a number of key players. Related to this (but more of an opportunity than a problem) is the fact that most virtual technologies make capturing knowledge and archiving work easier than in face-to-face firms. When all of a team's interaction is captured in e-mail, shared documents online and threaded discussions, there is already a tremendous base of knowledge and rich history of the team. The challenge then becomes how to identify the value and richness of all that information and find a way to archive it so that it becomes accessible.

Why is this important for a virtual team? If you operate under the assumption that there will be some turnover within the team, then archiving important information is an essential way to help prepare personnel new to the team. Additionally, even without turnover, accurate information (about what was said, what was agreed to or proposed, ideas discussed, options considered, commitments individuals made and outcomes achieved) from team interaction or individual work provides a significant advantage to the team. The next time there is a disagreement about what was agreed to, the team doesn't even have to consult the minutes (which may have been created sometime after the discussion anyway and are thus limited by issues of memory). They can pull up the archived e-mail or chat transcript in which the specific option was proposed at that time. Instead of having to type up a report to a client based upon a discussion the team members had, they can cut and paste the text from the groupware exchange and use that text as the foundation for the report. Right now, information-rich technology has tended to inundate professionals with information overload. Used wisely, this technology can enhance learning and improve knowledge management. But this will typically require at least one person with at least a part-time role of knowledge management responsibility and the focus of the team on information management issues.

Managing this effectively may not mean that a particular individual must be identified as a full-time archivist or knowledge manager for the virtual team. Some technologies can be configured to capture this information automatically. But at a minimum, some sort of protocol needs to be established to facilitate or support the capturing of key data. Otherwise, each team member will use different templates, categorize information differently and set up a variety of folders, and there will not be a common architecture that makes the data accessible. Imagine a library in which each individual librarian catalogued books according to their own system and set of guides. Thus you'd have a library with books stacked according to one of 12 different cataloguing systems. The only way to find items in such a library would be to wander through the stacks. A team without alignment on how to share, capture and archive data (as well as what to capture and archive) is the equivalent of such a library – a mass of interesting data that is basically inaccessible.

Another key function or role related to knowledge management is also skill development. Even at the initial start-up of the team, team members are (in terms of technological literacy and competency) likely to be all over the map. Even when team members come from the same organization, some individuals will be very familiar with the groupware or videoconferencing hardware the team will use and others will have shied away from it. It is important that the team members start off as close together as possible in terms of technological competency. That means that the team needs to find ways to either provide skills for the 'Luddites' on the team so they can participate like everyone else from the start. Or the team must provide some sort of skill-building resource so people can quickly close that gap. Thus one key element of every successful virtual team is a resource (a person, a help-desk, a FAQ file, a self-paced tutorial, a job aid, a buddy system – the possibilities are almost endless) that provides the technological skill-building and support for those team members who need it.

I am not arguing that any successful virtual team has a trainer dedicated solely to that team. Or that all team members must attend a course on 'effective e-mail usage' before the team can start up. Technological competency can be developed in a lot of different ways. Look at your own example. How did you learn to use a computer mouse? Did you have a personal tutor? Did you take a class that specifically taught you how to use a mouse? Did you sit down and read the computer support documents from cover to cover? If you're like most people, what you did was, first, fool around on the computer to see how things worked by trial and error and then, second, you played the computer version of solitaire (or some other basic computer game that used a mouse). Computer games have served as one of the best tutorial methods for instruction on using the mouse. There is a lesson here for learning and technology. Learning must be built into the technology so that eventually the hardware or software becomes natural.

There is an important point to be learned from this example. Some of the best learning does not occur in classrooms or in formalized learning situations. It can occur with fun and games. Or it can occur in situations that are structured to provide learning in 'baby steps' or small incremental stages while someone is doing work. For instance, when I facilitate a groupware session (often on an organization's strategic plan or brainstorming marketing ideas), one of the first ways I make sure everyone is conversant with the software is to have the participants use the groupware to determine where we're going to have lunch during the break. We might start by identifying criteria for a good place to eat (within walking distance, inexpensive, not a long wait to be seated and so on). We then vote on favourite cuisines. From there we brainstorm restaurants – what is in the area? We apply the restaurants to the criteria we've developed and rate them. After we're finished, people have acquired enough competency to participate in a more intellectually rigorous topic at a faster speed, they've lost any intimidation they may have originally felt from the software and they have also enjoyed themselves. Plus we've decided where we're having lunch (it's always essential to deal with the most important matters first when doing strategic planning!).

This groupware example applies equally as well to other technologies. Suppose that the team has a website just for the team to share documents, archive information, establish a threaded discussion and serve as a 'face' to the world as well as a point-of-contact within the firm. You can start by asking each member to go to the website and input some basic information (a picture of themselves, a short biography, what they believe they bring to the team). Activities such as this can easily be turned into team-building pieces where members are asked to share information about what expectations they have of the team, what their core values are, their previous experience with virtual teams, or what they think their team-mates should know about them. What is nice about such virtual activities is that they have multiple agendas. They build up an individual's competence with a particular technology and their confidence, so they can interact successfully using the particular tool. They can be fun or social so there is little pressure, mistakes are tolerated and patience is greater than it would be in a purely task-related situation. They can also be effective team-building activities that contribute to team trust and cohesion. Last of all, they are also good assessment pieces for team leaders and managers. If some members can't sign on to the website or don't complete the online task (as simple as it is), that is a quick message to the team leader that a particular member requires some individual attention to determine if this is a motivation or competency problem.

The key lesson from this example (of building technology training into initial activities) is that not only do virtual teams need to provide technology training for members to ensure minimum standards of proficiency, but that this training works best when it is combined with other activities. In the end, team members gain skills, but the training didn't seem like training. Instead, it was a game or a team-building activity or an initial task (like posting biographical information). Perhaps the best way to phrase this is: don't train – play!

Questions to consider

1. What process and behavioural expectations would you have for any virtual team and its members that you're a member of? What do you believe would constitute 'professional' behaviour from a virtual perspective?

2. Think of a project or team you're involved with currently at work (even if it isn't virtual). What would a good charter for that team look like – especially if that team were required to work virtually?

3. Think of ways to use games, social activities, start-up work and orientations as a means of technology training to bring new team members up to sufficient technological proficiency. How can you build training into these events or activities?

Notes

1. Charles Handy has contributed much to the understanding of organizations and the evolution of work. But I disagree with his thoughts on virtual work. See Charles Handy's comments in the *Harvard Business Review* on May–June 1995 where he wrote 'a shared commitment still requires personal contact to make it real … the more virtual an organization becomes, the more its people need to meet in person.' I think Handy is guilty of assuming that only some elements (such as trust) can be developed through face-to-face contact and that only face-to-face contact is capable of providing 'high touch' (a term coined by John Naisbitt in his book *Megatrends* (1982), meaning that the more technology-based our interaction becomes, the greater the need we have for non-technological and personalized – or 'high touch' – exchanges).

2. The leading researchers in this area have been Jarvenpaa and Leidner (1998).

3. Recent data about e-mail usage within the UK shows little agreement about what behaviour is appropriate regarding the use of company e-mail for personal uses. A 2002 study by Experian in the UK identified a number of unprofessional

e-mail tendencies. According to Experian, 60 per cent of employees in the UK admit to reading personal e-mail while on the job, 33 per cent had sent embarrassing e-mail to the wrong person, 30 per cent admitted to having a cyber relationship with another professional they've never met in person, over 30 per cent had asked someone out for a date using e-mail and 60 per cent admit to sending e-mail to a co-worker seated next to them.

CHAPTER 5

Hang together or hang separately

During the struggle of the American colonies to gain independence from Great Britain, American colonist Benjamin Franklin lamented the lack of trust among the colonies and their unwillingness to band together. Because the colonies insisted upon autonomy even from each other and were uncomfortable with any kind of central authority, establishment of a central military force and overall leadership to fight the British was stymied. It was at that point that Benjamin Franklin pointed out to his compatriots the need to collaborate as a group or face the consequences of failing in the rebellion and being tried for treason against the Crown. 'We must all hang together or assuredly we shall all hang separately,' noted Franklin.

Metaphorically speaking, the same is true of any work team. While most teams may not literally face the gallows as a punishment for failure, in today's organizations where layoffs are common and poor performance can shut down a business in this extremely competitive world, there is often a severe price to be paid for teams that fail. If nothing else, there is usually a personal cost to being part of an unsuccessful team. Being part of a team that fails can derail a career that is on the 'fast track'. At the least, participating in a failed team can be an experience that contributes to professional burnout and a great source of personal frustration.

The value of cohesion

A key element in almost all successful and high-performing teams is that of cohesion. Cohesive teams are those that don't fall apart under pressure or tough times. Instead of resorting to finger pointing and the 'blame game', when the going gets tough they perform. Typically the way that an uncohesive team reacts to pressure (such as a difficult assignment, some frustration, great uncertainty or widespread conflict among members) is with disintegration. Members drop out from the team (or if they can't quit the team, they work less and place lower priority on team tasks). There is less energy within the team and cliques begin to develop (or if they already exist, then they become more pervasive). With the cliques, distrust begins to grow – members perceive hidden agendas behind all actions (even when they don't exist). With the perceived existence of political agendas, leadership is seen as playing favourites, cynicism becomes widespread and everyone feels isolated. Political gamesmanship occurs as people assume all stated motives are false. Members of such teams are quicker to disown results they disagree with and the team becomes a source of conflict rather than one of strength. Teams that lack cohesion are ones that tend to give up when confronted with a barrier or difficult task or unexpected obstacle. Given that most assignments in today's world usually involve barriers, some degree of difficulty and unexpected obstacles, cohesion sounds like a very important thing for teams to possess.

Under levels of pressure that has uncohesive teams cracking, cohesive teams buckle down and work to get things turned around. Members of cohesive teams may not enjoy pressure but they don't disintegrate under it. The cohesion within the team means that team members trust each other to perform. Members of such teams focus on their own assignments to the best of their ability (rather than trying to determine if their mates have quit and left them on their own).

In today's work environment, there will be many temptations for team members to physically or mentally quit on a team. Whether this means actually leaving the team, doing shoddy work, putting the team as a last priority, or withdrawing from the team and its work by any of the members, quitting the team is deadly. Yet this is exactly what is likely to happen when a team lacks cohesion and strong bonds. In short, cohesion is an important aspect of a successful and high-performing team.

Building cohesion

How does one get a cohesive team? There are a number of methods for building cohesion within teams. Typically, teams that are very successful, with good performance records and confidence in their ability to manage difficult times are also very cohesive. In this instance, success tends to breed cohesion. Successful, high-performing teams create team atmospheres where members can tolerate small issues and own up to mistakes because the team is so successful. But this is of little help to us – it does not tell managers how to build a cohesive team (other than to find a team that will be successful and become confident – if only it were that easy!)

There are other factors that contribute to team cohesion. Team members that share common values and goals contribute to building team cohesion. Unfortunately, it is difficult in most organizations to identify and select personnel on the basis of personal affinity and strongly shared values. Elite military organizations (in which people may go through extensive selection trials and years of training) can sometimes achieve cohesion through this manner. Another method for building cohesion is through life-altering events. Individuals that go through shared strong, traumatic experiences often bond. For instance, people that live through natural disasters or wars may develop strong ties to each other despite interpersonal or class differences. Thus many organizations attempt what is called 'challenge technology' as a form of team-building (such as rock climbing or white-water rafting). However, one of the limitations to this as a method for building cohesion is that it can just as easily tear a team apart (especially if some members fail to perform or the team fails – in which case members may question the value of belonging to a team with a group of 'losers'). Additionally, it is just as likely that a significant challenge (capable of building cohesion if the team can master it) may also tear a team down if it fails to succeed. After all, these events serve to bond people to the extent that they are serious challenges (which means there is a realistic chance of failure and the stakes are high).

The most important element in creating a cohesive team is that of trust. A team where members trust each other is a team that is cohesive. Where there is trust, there is confidence and a degree of affinity. Teams in which members trust each other are more likely to bear up well under pressure – because they believe their team members aren't operating with hidden agendas and can be counted on to do what they say and perform well. Teams with high levels of trust also display high degrees of cohesion. Consequently, creating trust is the most effective means of creating team cohesion.

Understanding trust

What do we mean with the word 'trust'? Professional literature and academic research provides us with hundreds of definitions of trust, all with very laudable and credible justification for their distinctions. You may have your own definition of what constitutes trust. However, for our purposes, we can define trust in a team as: a belief that people will do what they say they will do and a belief that the team and its members are competent performers. This is not meant to exclude other interpretations of what trust is or to argue this definition is superior – only to provide some clarity about what I mean when I refer to the concept of 'trust'.

These points about trust and cohesion are important because not only do trust and cohesion matter to the success of all teams, but there are a number of incorrect (yet widely held) assumptions about trust and virtual work settings. It is a widely popular belief that teams (or workforces) that are totally or mostly virtual are incapable or unlikely to develop trust.[1] Even a number of reasonably informed and knowledgeable authors on the subject of virtual work argue that for a virtual team to succeed, it must first start with a face-to-face meeting in order to build trust. Such assumptions have major implications for virtual work settings because if they're true, they seem to indicate that virtual workforces are doomed to fail (because they won't ever develop trust and trust is vital for serious work, especially within teams, or because they'll always be inferior to their face-to-face counterparts). And, increasingly, there are too many virtual teams who never have opportunities for even one face-to-face meeting, let along an extended initial team-building session to help kick-off the work. What we have found is that these assumptions are wrong. When one views trust as something that is developed in a face-to-face setting, it is logical to then conclude that interaction without face-to-face interaction would therefore be devoid of trust. But what we have discovered is that trust develops in a number of different ways and there are a number of different levels of trust. Consequently, it is wrong to believe that trust may only be developed from face-to-face interaction. Purely from an anecdotal perspective, we see numerous cases now where committed spouses with children and an apparently happy marriage abandon it all for life with someone they have never met face-to-face but have interacted with only online. These instances clearly demonstrate that trust can be developed virtually. So we must now try to understand how it develops, in what circumstances and the application to work situations.

If there are hundreds of possible definitions of trust, there are just as many models explaining levels of trust or how trust progresses.

One particular approach that has particular appeal (especially because of its simplicity as a model) explains the development of trust in three stages.[2] The initial stage of trust is what might be called reciprocity or punishment. I can trust you to do what you say you'll do because if you don't, I (as your supervisor) will give you a bad evaluation or perhaps even get you fired. As negative as this sounds, we see this form of trust displayed in modern society on a continual basis. When you choose to eat at a new restaurant or take a commercial flight on business, you trust the cook to prepare the food correctly and the airline pilot to operate the aircraft safely. You trust each of them to do so even though you probably have no idea who the cook and pilot are and have never seen a résumé or safety record for either of them. The primary reason you trust them isn't that you're naive. Instead, you know that if the kitchen prepares the food poorly you'll complain (and the restaurant owners will replace the cook if enough complaints come in). If the restaurant doesn't follow sanitary standards, you can get it shut down for health code violations. If the pilot operates the plane unsafely, they'll likely be grounded. Any actions on their part that endanger you are just as likely to endanger themselves. Thus in modern society we place complete (well, usually complete) trust in total strangers who drive our buses, fly our planes, deliver our mail, prepare our food, conduct surgical operations and process our credit cards because we know that if they violate our trust they will have a price to pay.

This progresses to the second level of trust. We can refer to this level as habit or track record or knowledge-based trust. In other words, if you and I have worked together for ten years and you've always arrived at work on time, when you promise to arrive on time for tomorrow's meeting, I trust you to do so. Why? Because you have always done so in the past and I expect you to do so in the future. I trust you to do what you say because you've always done so in the past. I feel like I know you well enough to be able to predict what you will do in certain circumstances. Thus this level of trust is defined by your previous actions. While I might say that 'I trust you', what I'm really saying in this instance is that I expect or predict you'll show up on time for the meeting because you're timely. In this case, trust is a function of your past behaviour.

Obviously, one's past behaviour does not always predict one's future behaviour (to paraphrase many investment firms and their commercials). However, when it comes to trust, if your word has been consistent with your actions previously, I am likely to trust your word now. Therefore this form of trust occurs only when someone has had an opportunity to demonstrate repeatedly that they have been worthy of trust in the past through their consistent behaviour.

The third and final level of trust in this model is one that is based upon values and perceived affinity. We can refer to this as the mutual identity or identification-based trust stage. In short, individuals at this level of trust (which is the strongest of the three levels) trust others because they believe they share key common values, beliefs, concerns or priorities. They have come to identify with each other. Though you and I may be very different in many different ways, I may feel that on the key issues, we're totally aligned.

Take the earlier example of whether or not you trust me to arrive at a meeting on time. Initially, you trust me to make the meeting time because you can get me fired if I show up late. Later (over a period of time as we work together), you come to realize that I'm hopelessly anal-retentive and always on time – so you trust me because you believe it is my habit to always be on time. But a meeting comes along that you feel is crucial to the health of our organization – and you perceive that I feel likewise. Now you trust me to show up on time for the meeting primarily because you believe that I think this meeting is vital and possibly the most important matter I'll face this month. While this might demonstrate that you're extremely naive (as I actually have a very cavalier attitude towards meeting times or I've become bored with work and have decided to start showing up late), the degree of trust you place in me will be greater if you perceive we share values than if you trust me because of habit (I usually show up on time) or reciprocity/coercion (if I don't show up on time, I'll face serious consequences).

Trust and virtual work

The research in the field of virtual work is challenging in a number of ways. For starters, it is difficult for researchers and academicians to stay current with existing organizational practices. As much virtual work has been driven out of necessity, organizations that do a lot of virtual work tend to be more focused on getting results and moving forward than really trying to understand why it works or what implications this has for other organizations. Consequently, many of the real insights around virtual work have been uncollected as organizations attempt to cope with time pressure and rapid change instead of collecting insights into processes.

Only recently has research begun to focus specifically on virtual teams and various group dynamics. From this research we have found some fascinating (well, fascinating to me at least) insights. For instance, one term that has been coined by researchers is that of 'swift trust'.[3] This term refers to a phenomenon in which workforces or teams develop trust very quickly. The research on 'swift trust' has identified that this occurs when teams tend to focus on key task matters and then use the trust that has developed to build member trust in other areas. In other words, we start with trust in each other's professional competence (our technical skills and ability to get things done) and use it to build and create trust in each other's agendas, ethics, openness and honesty. This approach to trust is interesting because it focuses more on the third level we discussed (shared values and affinity) than it does on deterrence and habit.

Research with virtual teams has found that they are capable of developing 'swift trust' without any face-to-face interaction.[4] In fact, some researchers have discovered that virtual teams are capable of developing trust more quickly than face-to-face teams. Interestingly, the initial research on 'swift trust' viewed this type of trust as somewhat superficial and useful as a launching pad for producing a stronger, deeper kind of trust. But virtual teams appear capable (because of the dynamics of virtual work and the different challenges) of not only producing 'swift trust', but also experiencing a much stronger degree of trust from this form of conviction and generate levels of trust that are stronger and deeper than their face-to-face counterparts. To translate this into plain language, virtual teams can develop trust faster and (in some instances) display stronger levels of trust than traditional teams.

How is this so? There are a variety of theories to explain this phenomenon. What appears to happen with virtual teams (if you remember our earlier model of the three levels or stages of trust) is that virtual teams tend to skip the first two levels of trust (reciprocity and habit) and jump straight to the third level (values). Consequently, virtual teams are capable of achieving trust very quickly. Additionally, if virtual teams develop trust by building on affinity or shared values, this degree of trust is likely to be stronger and be able to sustain stress better than trust dependent upon reciprocity or habit. Thus some virtual teams appear to manifest deeper and stronger degrees of trust than their face-to-face counterparts.

There are other important consequences for virtual work. Apparently, virtual teams tend to develop trust quickly – or not at all. A face-to-face team may muddle along, displaying hidden agendas, infighting and distrust only to eventually make a quantum leap forward through some trust-building event. A virtual team, however, either develops trust relatively early in its existence or doesn't appear to ever develop it. For managers, this offers a key insight – monitor the development of virtual teams in their early stages. If the team's task involves one in which trust (or collaboration or some other behaviour that requires trust) must be present, any virtual team that fails to develop this attribute early is not going to produce it later and should be terminated. Another implication for managers is that if a task requires a high degree of trust and cohesion, a virtual team may be a better option (because it is capable of developing trust faster and at potentially stronger levels than a more traditional set-up). Consequently, managers may deliberately choose virtual set-ups over face-to-face ones because of the need for strong and quick trust required by the nature of that work.

Building trust in virtual environments

While virtual teams are capable of developing 'swift trust' this does not mean all teams do so and that it easily occurs. There are a number of factors which encourage the development of trust in virtual teams (as well as elements that discourage the development of trust). Ironically, many of the predictors of trust in face-to-face settings appear less relevant in predicting the development of trust in virtual settings. This is probably due to the difference in dynamics of virtual and face-to-face settings. Also, it appears that task clarity, focus and productivity appear to play major roles in the development of trust virtually. Consequently, issues like 'eye contact' (which matters a great deal in North America – where people are often assumed to be shifty and untrustworthy if they don't make eye contact) appear to be de-emphasised or irrelevant in virtual settings (because we tend to focus less on behaviour and more on performance). This is important to note because this may explain why some theorists had initially claimed that trust is impossible in virtual settings. After all, if non-verbal behaviour is a prerequisite for the development of trust in traditional settings, then a virtual work arrangement without visual contact would appear to make development of trust impossible. Except that in virtual settings, trust develops differently and the criteria that people appear to use for deciding if they trust or don't trust someone are different in virtual settings.

Typically, virtual teams will fail to develop trust if one of several factors occurs. First, if the exchange or interaction between team members is sporadic or ineffective, trust won't develop. Professor Catherine Cramton at George Mason University found that perceptions about response quality and timeliness shape how people regard their team-mates and how much they trust them.[5] Cramton quoted one member of a team she studied as saying, 'Everytime I sent an e-mail requesting "any thoughts?" I expected to receive one from everyone. And when I didn't, I felt they weren't holding up their end of the bargain.'[6] Thus vague or general language or different standards for response time may prevent trust from developing within the team.

Strong leadership is also a powerful factor in virtual trust building. On its face, this may sound like an obvious position – what team or organization wouldn't benefit from strong leadership? But from a virtual perspective, 'strong leadership' means that the manager or team leader needs to demonstrate more authority and direction than they would in a traditional, face-to-face setting. There are many times in face-to-face settings where a manager or team leader may withhold their own viewpoints to encourage openness and free expression by team members – especially initially when team norms are being formed. This approach is often very effective in building trust, openness and collaborative communication. In a virtual setting, this behaviour may be counterproductive. Leaders need to be stronger and initially more directive in virtual settings – not because team members refuse to take initiative, but because strong initial direction is reassuring to team members. Rather than cutting off discussion, strong directive leadership early in the virtual team's birth serves to build confidence and trust among the team – reinforcing the belief that the purpose is clear and confusion won't reign.[7] As consultant and author Trina Hoefling noted about her virtual team research, 'people want to be reassured they aren't lost in cyberspace.'[8] Strong direction by the manager or team leader serves to reassure virtual team members. Conversely, they are also less likely to be intimidated by a strong leader in virtual settings. These points are important to make because in face-to-face settings, we might encourage leaders to be less dominant and allow the group to express opinions and help set agendas – to let members know their input is welcome. However, in a virtual setting, a strong leader is important for the development of the team yet doesn't appear to dominate membership and stifle team development like it does in face-to-face settings.

Another factor that prevents trust from occurring within a virtual team is when team members perceive there is a lack of competence or initiative.[9] This is an important insight because there are plenty of examples of face-to-face groups where members trust each other and there is no initiative or the team may not accomplish anything. But virtual teams appear to be more focused on performance as a prerequisite for trust. If the team encounters initial technological barriers (some members can't receive e-mail attachments, the server is down or the sound for the videoconference isn't working) and the team fails to adapt, then members unconsciously tend to view this not as a technical failure but as a verdict on the team. Consequently, it is important for initial team meetings to 'work well' or for individuals to jump in and solve problems. For instance, if the 'chat room' isn't working, the team leader (or another member) needs to quickly suggest 'let's just conference call instead since we can't use the chat room' or 'OK, we'll go asynchronous – everyone trade e-mail on this issue and let me know what you think.' The ability to adapt – particularly to technical obstacles – is a key factor in building (or the failure to build) trust in a virtual team.

Contrast that group dynamic to those in a more traditional team. If you went with a group of co-workers to meet in the company conference room only to discover the CEO had taken it for a special meeting, would you and your peers just throw up your hands and give up? Of course not – you'd find some other place to meet (even if it meant a hallway or pulling some chairs into an individual office or going to a coffee shop). And doing so wouldn't usually make your face-to-face team more trusting – traditional teams and groups would simply see this behaviour as adaptability. Yet in a virtual setting, it is just such adaptability that builds trust in virtual settings. Perhaps this is because trust in a virtual environment is more of a factor of professional competence and confidence in the mission and members. In any case, activities that don't contribute much to trust in a face-to-face setting do build trust virtually. Consequently, some of the behaviours that contribute greatly to trust in face-to-face settings have less impact as trust-building measures in virtual settings and ways to build trust virtually have less impact for traditional settings.

Strategies and activities

Anyone planning on creating or starting a virtual team or work unit should carefully consider how he/she intends to foster some degree of trust within the group. There are several strategies that are likely to maximize the likelihood that trust will develop (and develop quickly) in that virtual work unit.

- Initially, make sure that team interaction has a 'high process' component. Focus much of the initial team interaction on personal information (sharing details about each other) and process (how the team will interact with each other) with less initial focus on the task and work details. Don't worry – successful virtual teams aren't primarily social support networks or party animals. Instead, what some research about virtual teams shows is that initial conversation is very social and process-oriented and then after a few meetings becomes almost entirely task-oriented. It's as if the team has to build up a certain degree of trust and comfort and orientation with each other – and then they focus completely on getting the work done!

- Strong directive leadership is obviously important initially. The manager or team leader doesn't need to be a dictator and, as the team grows, they can be far more laissez-faire or consensus-focused. But initially, the group members need to come away from their first team interaction assured that the team has a clear sense of purpose and goal, that they are not 'wandering in the wilderness' or 'lost in space'. A leader who works in both virtual and face-to-face environments should plan on being more directive and 'in charge' in virtual settings.

- The team needs to demonstrate quick initiative – especially towards technical problems. This builds trust within virtual teams because members then perceive everyone else to be competent and committed (rather than confused and hesitant). It builds confidence in the rest of the team. This attribute is important enough (and also so likely to come up as a challenge as nearly all virtual teams will have some hardware or software connection problem) that some savvy team leaders will build in deliberate 'breakdowns' that a team member has been briefed to solve as a means of building member trust ('wow – look how quickly Nigel got the network back online!').

- Quick responses by team members to queries and requests builds trust. Thus teams need to either start out with very eager team members or the initial team interaction needs to examine assumptions about response time. The team leader and members also need to do what they can to encourage the 'energy level' within the team. This also makes the case for both posing some initial questions that you know members can answer and also 'seeding the field' by individually encouraging some members to respond quickly (and thus set an example for the rest of the group). With lots of virtual technology, this is done very easily. Many chat functions also allow for private messaging. Thus you can pose a question to the entire group but then send a private message to Doug indicating he should set an example for the group and show everyone how much he knows on this subject. I've done the same with listserv and e-mail (where the listserv request goes out to all members but I then e-mail individuals to prompt them to respond). After the first few queries, the tone is set for the group and the perception is created that this team is a bunch of hard-charging, super-competent, energized, top performers. And this generates trust within a virtual team.

- Frequency of interaction is very important. Generally speaking, the more interaction, the more likely the team is to build high levels of trust. It is important to note that frequent interaction is not a sufficient predictor of trust. This is also not to say that a lot of spam and junk messages is a good thing. However, high communication levels appear to reinforce and build upon established trust. Related to frequency, predictability is also important. Team members have shared expectations about how quickly people will respond and what kinds of message require a response as opposed to those for which a response is optional. As a rule, I encourage face-to-face teams to formally meet only when they have business to discuss. After all, if you pass people in the hallway on a regular basis, why meet every Monday afternoon unless you have real business that needs to be dealt with? But with virtual teams, it's incredibly easy for silos to develop. So I encourage virtual work units to set up regular times to 'check in' and 'touch base'. Thus rather than schedule conference calls or online meetings for virtual teams based upon a need to conduct business, recognize that a key element to a successful virtual workgroup is the frequency and predictability of interaction. Facilitate this frequency and predictability by scheduling regular sessions for the team to interact.

- Perceived performance and professional competence are vital. Simply put, most staff who are asked to be part of a new virtual team are silently asking themselves the question, 'Is this a waste of my time or are people here going to produce?' Thus information at the outset that informs team members about their peers' credentials and competence, the identification of clear performance measures, the celebration of individual or team milestones – these are all pieces that contribute to and maintain trust within virtual settings. As mentioned previously, opportunities that allow individuals to shine or demonstrate initiative also demonstrate competence and build trust.

There are a number of specific team-building measures managers or facilitators can take for new virtual teams to enhance the likelihood that trust will be developed within the group.

- Establish a team biography page on the intranet or team website or threaded discussion. Have each team member provide information that provides a sense of professional competence and who they are as a person (include pictures too). Generally, this does not mean just asking people to post a curriculum vitae. Instead, you may wish to ask a series of questions (which team members answer). I've seen some instances where the questions range from the professional (what obstacles did you encounter in your last project and how did you deal with them?) to the fanciful (if you were stranded in an elevator for three hours, who would you want to have in there with you and why?).

- When the team interacts virtually, have each member share what is happening in their respective location. This is especially powerful if the team is distributed across different time zones or continents. Creating a visual picture of the environment of each of the team members plays a surprisingly strong impact in building a sense of a great whole (rather than a collection of individuals). At subsequent team meetings (conference calls or chat sessions), give team members a chance to share what has happened to them in the time since the last team meeting. You want to help members see how they're all evolving – to paint a picture of each individual changing over time – rather than a static one of a list of credentials and accomplishments.

- Set up an online 'café' or parking lot where team members are encouraged to socialize and focus on non-task-related interaction. This approach works especially well with organizations or large teams and interaction involving websites. I've seen some organizations even set up 'amnesty' rooms where all comments stay in that chat room, rank disappears and no one can be disciplined for what they say in the room. There are a range of different approaches you can take but the key lesson here is the value in officially creating meeting space for team members to vent, blow off steam, socialize, play or be off-task and do so with permission.

- Look for opportunities to create norms of initiative, self-reliance and shape perceptions of professional competence. This might mean an initial activity in which team members are paired up (almost in a 'buddy' or mentor relationship) and given an initial assignment that helps acclimatize them to their new virtual environment. This could involve a scavenger hunt for those interested in a more social focus (looking for particular items on the company website). Or, a more process-focused activity might involve pairing members up to develop initial recommendations on particular process issues (team ground rules for virtual meetings, how to archive team communications, a list of member expectations for how the team should operate, identifying areas where team members require skill development or training to master particular hardware/software required for virtual interaction). The idea behind all of these activities is to provide an opportunity for team members to experience others on the team demonstrating energy, initiative and professional competence – even in small ways. By doing so, this builds confidence in the virtual work unit and, more importantly, trust in each other.

What would you do?

> You are the team leader of a virtual workgroup that has already demonstrated strong performance and appears to have developed swift trust. The team members act quickly and seem to respect each other. As team performance increases, each person is taking on greater workloads. The consequence of this is that team members appear to have less interaction. Additionally, you've recently added a new team member – Tess. Tess is known by several of the veteran team members and respected by them. She has a good record professionally. Unfortunately, Tess hasn't performed well so far – turning in some assignments that were done poorly or failed to meet deadlines. What should you be on the alert for so that team trust doesn't drop and team performance continues to accelerate?

Well, because you're a savvy virtual team leader, you already know that one of the things you want to do to maintain trust is to continue with predictable communication. Even though the team performance is accelerating (and, as a consequence, team members are spending less time communicating), you will need to continue to emphasize this issue (predictable communication). As tempting as it is for hardworking team members, they can't put off responding to queries or online requests – they need to respond quickly and consistently. As the team leader, you'll need to keep an eye out for this – and monitor the communication habits of the team.

Tess (the new underperforming team member) creates a unique challenge for you. Some research on virtual teams indicates that when managers or leaders are seen providing negative posturing or going outside the team with negative information, trust suffers.[10] For instance, in a face-to-face situation, if your manager asked you 'how is Tess working out?' everyone on the team would recognize it would be appropriate for you to tell your manager about the problems you're having with her. But in a virtual setting, that exact same response on your part is more likely to be perceived as 'airing dirty laundry' or 'being a tattle-tail' with adverse consequences for team trust. Thus you're going to need to tread lightly on the issue of what performance problems within the team you share outside the team. Ideally, you'll request for (and get) lots of freedom from your own manager (so you have time to resolve performance issues first before you need to report out that it isn't working). Additionally, you'll want to try and make Tess's performance assessment as objective as possible. Therefore focus on outcomes or results, not behaviour or attributes.

Questions to consider

As you think about trust and cohesion in virtual work environments, consider the following questions about your own experiences and expectations:

1. Think about virtual work environments you've observed. What were the differences between the high-trust environments and the low-trust situations?

2. Does your current or future virtual work setting benefit from or require high levels of trust? If so, what could you (or others) do to enhance the development of trust within the group?

Notes

1. Unfortunately, there are a number of very well-respected and useful sources on virtual teams that insist that the most effective means of building a virtual team is to start with a face-to-face session. Lipnack and Stamps (1997), and Duarte and Snyder (2000) are two excellent and widely read books on virtual teams that both insist an initial face-to-face meeting is necessary for building trust. Many other researchers (Nohria and Eccles, 1992; O'Hara-Devereaux and Johansen, 1994) have insisted on this as well.

2. From the *Harvard Business Review* in May/June 1998, pp. 20–1.

3. The term 'swift trust' was first coined by Meyerson, Weick and Kramer (1996).

4. Jarvenpaa and Leidner (1998).

5. Catherine Cramton has done extensive research in virtual work and communication. This is from 'Finding common ground in dispersed collaboration' in *Organizational Dynamics* (2002).

6. Cramton (2002).

7. Jarvenpaa and Leidner (1998).

8. Hoefling (2001), p. 43.

9. Jarvenpaa and Leidner (1998).

10. Jarvenpaa and Leidner (1998).

CHAPTER 6

Managing – a brand new day!

'Try not. Do or do not. There is no try.'

Yoda, from *The Empire Strikes Back*.

Being a manager in the organization of today has involved a series of revolutions. Managers must now deal with far more diverse workforces with multiple cultures and languages than yesteryear's managers. The increasing use of teams in organizations has changed the role and skillset necessary for an effective manager. Today's managers now face heavier workloads, a larger span of control, greater time pressure and more complexity than their earlier counterparts. They face less stability and more ongoing change. Managers are more likely to work in organizations that are much more fluid in structure and culture – forming alliances with competitors, merging with others, moving staff to contractor status and undergoing a constantly shifting sequence of form and function. Yet virtual work will create more changes for managers than what they have faced to this point. The realities of virtual work will force management to undergo a massive transformation.

Out of control

Let's examine some of the ways that virtual work and virtual teams will change the way effective managers manage. The widespread reliance on face-to-face teams in many organizations has already forced managers to adapt. Teams have forced managers to focus more on building consensus, delegating, pushing authority downward and serving as an enabler for the team – fending off interference from the organization to allow the team members to focus on their work. But the development of virtual teaming has resulted in more change for managers. In the past, most supervisors managed within line of sight of their personnel. While managers might engage in some travel or attend meetings that meant work by staff sometimes occurred without a manager present, such cases were the exception and not the norm. Managers have been taught (and have grown accustomed to the belief) that managing people usually required that they be within sight. Or at least that with a little travel (a walk down a hallway or across a street) a manager would be face-to-face with their employees. This is because we have encouraged managers to manage by dealing with employee behaviour. Most managers tend to focus on behaviour with their staff (looking for hustle, positive attitudes, initiative, hard work, good work habits and a host of other positive attributes). Ask most managers to describe their best employees and they'll talk about them in terms of their behaviour – what it is that those staff members do and say and think and behave. To manage behaviour means to be able to see that behaviour. Thus we have a whole cadre of managers and supervisors who are used to observing employees. This ability to watch people work is conducive to an atmosphere of control. Indeed, those who believe that workers cannot be trusted (while the cat's away, the mice will play) and must be controlled are therefore comfortable with a supervisory set-up that allows a manager the opportunity to watch staff work on the job.

But with virtual work settings, managing by observing staff is often impossible. Let's take the examples of dispersed or hotelled workforces. Hotelling involves mobile workers who are typically on the road using temporary space or other locations. With dispersed and hotelled staff, it isn't geographically possible for a manager to directly observe staff because they are too spread out for one person (or even a manager and several assistants) to be able to travel and watch them enough. Workforces that allow telework and telecommuting (where employees may work from home or work on flexible work schedules) make visual control nearly impossible because it isn't realistic to expect a manager to travel to one's home to observe someone in action. And the percentage of workers who engage in telework or telecommuting is only going to increase (because of people's desire to avoid commutes and adopt flexible schedules, and the need of organizations to keep scarce talent that may be on maternity leave, tired of commuting or ready to retire).

Shift work also creates situations where managers would need to be at work 20 hours a day (seven days a week) to provide some degree of visual oversight of workers (some of whom may come in only on weekends or in the evenings). And as organizations become more global with their markets and products (and thus need to be open on the weekends because a weekend in London may be a weekday in Australia) need to adopt shift-work arrangements (in order to staff call centres, update websites and keep information current).

Additionally, organizations are more likely to use consultants and outsource various pieces of work – another trend in today's workplace. It's usually easier (and cheaper) to use a consultant (who already has expertise in an area) to do a specific piece of work than it is to advertise for, recruit, hire and then train an employee to do unique work the firm wouldn't normally do. Additionally, organizations may choose to outsource areas that are not regarded as key competencies for the firm. The outsourcing of work and reliance on consultants makes it difficult to directly observe their work because the consultants or contract workers will often perform from their own office.

In each of these cases (hotelling, telework, shift work, outsourcing) the typical manager finds themselves in a quandary. Having been taught that to manage effectively, they must control their personnel, how do you control someone you cannot see? Much of the skills and tactics and expectations we have encouraged or developed in managers are inconsistent with the reality of virtual work environments. Consequently, modern organizational life is, for some managers, a nightmare. Many managers may feel that the world they must work in and the work they must manage violates the rules they were taught.

Some of you who are sceptics are probably insisting that the new working arrangements don't prevent observation of workers.

What about supervisors or deputy managers at dispersed worksites who report back to the manager what it is they see and what behaviour they observe? And doesn't the new technology allow for possibilities like video surveillance or listening in on phone calls or auditing e-mail? If an organization truly insists upon observing the behaviour of those that work at the firm, there are indeed a wide array of options for obtaining that observation (some of them involving technology). If managers expect supervisors or on-site managers to observe and then report back, this still creates major dilemmas for any dispersed organization. This approach leads to additional bureaucracy (and the more layers within an organization, the more information is filtered). Thus putting supervisors at each job site (so that the single graphic designer or programmer or secretary located there has an on-site manager) actually reduces the amount of direct information to management by adding layers of personnel. Part of the reason many organizations got 'flatter' and eliminated layers of management in the 1980s and 1990s was to reduce costs. But an equally compelling reason was to improve efficiency and communication by 'delayering'. The approach of insisting upon direct supervision for each employee located away from their original manager may artificially create observation, but it simply isn't practical for most organizations because of the cost and reduced communication that occurs with this approach. Additionally, managers still find themselves in a situation where they must trust the reports of the supervisors who are on-site. And one of the reasons many managers resist off-site work arrangements is that they don't trust the information they receive – they feel they must observe behaviour and judge for themselves. Plus, managers tend to be so much busier in today's environment that even those who work at the same site as the staff they supervise discover they are spending too much time in meetings or travelling to do much direct observation. So locating supervisors at each worksite doesn't provide a solution for those who insist upon direct observation.

Technology may appear seductive with this challenge (attempting to observe offsite performance). There are anecdotal examples of parents leaving hidden video cameras to observe the nannie's behaviour around the children during the day or to catch thieves and vandals. But using technology to observe workers (whether it involves video cameras at off-site locations, monitoring phone calls or reviewing e-mail) is generally a failure. While there is a certain amount of fear and paranoia among workers that the new technology will be used to spy on them, the reality is much different. E-mail is one technology that is, in theory, easy to monitor. Almost all corporate e-mail systems go through company servers or computers, so management can access individual accounts and routine back-up procedures by the IT save most or all e-mail sent and received. Despite policies against using company e-mail for personal messages, a high percentage of employees in the UK admit to sending and receiving non-work messages[1] (one may assume that the numbers violating such policies are even higher than those who admit to this practice). In a modern organization, the amount of e-mail, phone mail and fax traffic is so great that attempting to monitor, collect, collate and read through such information is an overwhelming task.

Outcomes not behaviour

Such extreme efforts to observe behaviour (such as installing video cameras or phone monitoring systems) misses an even more important point – managers should focus less on behaviour and more on results. Ultimately it is results or performance that an organization is interested in. Stock prices for a company don't go up because the employees work long hours or show 'hustle', but they will go up if individual productivity improves. At the end of the year or the end of the quarter, the organization puts far more weight on results and outcomes than it does on activity. The argument here is not that the ends justify the means. Rather if the organization has goals, then the first and most relevant evaluation should be whether or not the goals were met (or progress was made to the goals). Evaluating behaviour then becomes relevant only to the extent that the goals weren't produced and we're certain that the behaviour in question contributes to whether or not the goal is achieved.

Take the simple example of a member of the company's sales force. Let's assume that one of your salesmen (whom we'll call Nigel) displays a number of behaviours that might irritate many managers. Nigel appears to daydream at times when he's at his desk. Nigel doesn't always respond to internal messages and memos promptly. His sales kit (with product information and brochures) isn't well organized. Nigel can often be observed wandering through the office, striking up conversations with lots of different people (most of whom aren't in sales). He's often late to staff meetings. And Nigel does a poor job filling out his timesheets and travel records (they are sometimes late, often with details scribbled out and written over in the margins – legible but not nice to look at). If we evaluate Nigel on his behaviour, clearly he's been found wanting and most sales departments would certainly have some employees who display the deficiencies of Nigel. And it's exactly the Nigels of the world that would have most managers insisting that is why it is so essential to be able to observe their staff – to keep an eye on them so they don't misbehave and do as they're told.

But Nigel also happens to sell more (both in number of sales and amount of money generated from each sale) than any of the other members of the sales force. This is not a fluke or accident. Nor am I arguing that talented people have personality disorders or that organizations should tolerate a 'star system' (where the top people are allowed excesses that other mortals are not). But if you look at behaviour, Nigel has a number of behaviours that would irritate most managers or that some would judge to be the sign of a poor worker. But if we look at results, Nigel has outcomes that demonstrate he's better at what he does than anyone else in the department. One of the messages Nigel's manager might get from this simple example is that the behaviours we initially focused on (with regard to Nigel) aren't relevant to his ability to do his job. We might as well evaluate Nigel on the basis of his shoe size, what side he parts his hair on or what magazines he subscribes to at home, because those appear to have as much relevance as the other behaviours we previously mentioned.

If we look at what performance we expect in order for someone to be a good salesperson (i.e. to sell a lot of our product) we would look at particular areas of performance. What is this person's 'close rate' (or ability to close a sale with an agreement by the client to buy)? What research does the salesperson have of their territory – how detailed and how recent is their information of prospective clients? Since you can't make a sale unless you meet with the client we'd want to know how many sales meetings our salesperson has set up with potential clients. We know that some sales generate more business than others so we might also want to look at the income produced per sale. Finally, good sales work often involves collaboration (such as sharing potential leads, working with other members of the sales force on proposals, telling administrative staff what they can do to support the sales force). So we might also look at the percentage of group proposals awarded or won that our salesperson had participated in (compared to the winning percentage of other group proposals) as a means of judging both the degree of collaboration with others as well as the success of that collaboration.

Using those criteria, this is what we would find out about Nigel. Nigel has the best close rate of any of the sales staff (a full 15 per cent higher than the next best performer). Nigel's records of his territory (list of potential and actual clients) are the most detailed and appear to be the most current. He also appears to have had more recent contact with his clients than any other salesperson. Nigel generates more total sales and also more 'big ticket' sales. And when he participates in group proposals, those proposals are 36 per cent more successful than group sales proposals that he doesn't participate in.

There should be at least two important themes in these sales elements we've just looked at. First, you'll note that none of what we listed under important evaluation factors for a salesperson were behaviours – they were outcomes. At the end of the day, when Nigel has gone home, we can tell what his close rate is or what his overall sales figures total because we're looking at outcomes and not behaviours. A result or outcome exists when the worker is gone. We can easily determine how complete and up to date his client territory information is. We can calculate what his close rate is or his sales volume is. The elements that are the most important for most salespeople (and, indeed, most professionals) are the outcomes they produce. The behaviours, in most cases, are important only to the extent they produce poor outcomes or violate the ethics and values of the organization.

Second, you might have noticed that none of the behaviours we found Nigel at fault for were listed here as important traits or performance indicators for a successful salesperson. It may irritate you if someone behaves in ways you find contrary or unprofessional, but is that behaviour relevant to their performance? It's easy to rationalize how it does matter. But if we looked at the original faults we had with Nigel (poorly organized sales materials, daydreaming at his desk, poor record-keeping and slow to deal with messages and requests), it's easy to rationalize how these are all signs of an ineffective salesman (if his sales kit is disorganized, he won't impress the client and won't be able to find material when he needs it). If we started by looking at the ultimate outcomes (sales) and then examine what performance is necessary to consistently produce sales, we did not list any of the behaviors that we originally were critical of Nigel for in this instance (sloppy travel reports, daydreaming, etc.). On the performance that matters for his job, Nigel is apparently outstanding.[2]

The first and most important shift that managers must make with virtual work is a shift from management of behaviour to one of management of outcomes or performance. Instead of seeking to observe how people act, managers in a virtual workplace will need to concentrate on measuring the results of what people do. While this is initially terrifying for many managers (because this is a significant shift for many managers in how they operate and an even bigger shift for many employees in how they work), this shift actually simplifies work.

As noted in our example of Nigel the salesman, performance or work outcomes persist after the employee is done or gone for the evening. Instead of focusing on what Nigel does, we focus on what he has achieved. If we set up some kind of consistent measures, it is actually easier to assess results than behaviour. For instance, what is easier to measure: Nigel's sales volume for the day or how friendly Nigel sounds on the phone during the day? Unless Nigel doesn't get any calls on the day in question, the result or outcome is easier to evaluate and measure. Generally speaking, assessing performance (meaning outcomes) will be less subjective than measuring behaviour. Consequently, performance appraisals tend to go better (because there is less dispute about the ultimate evaluation). Employees are more likely to know ahead of time how they're doing (because they can usually track their results while self-assessing behaviour is so subjective and also dependent upon what behaviour and when it is observed). And ultimately, it's the performance – the results – that the organization is interested in. Again, we're not claiming that the ends justify the means, only that focusing on behaviour is harder to do and less relevant than focusing on results. And for those managers who must operate in a virtual workplace, measuring results and focusing on output is not that difficult to do. In some ways, virtual interaction may actually make it easier to aggregate outputs (because information may go into a database or on a website directly).

This insight (about a shift from managing behaviours to managing results) has several practical implications for managers. Managers will need to develop new work standards for staff. In other words, what people are appraised on will need to be changed. This development of new work standards needs to be done collaboratively (so staff and management understand each other and what particular standards mean). These standards need to be output or results-based (rather than activity-based). A general test of this standard is whether or not we can judge the work of the performer after they are gone. For instance, if one form of evaluation is that an employee be a 'self-starter', how do we know if they were a self-starter after they left work? If, however, we state this job requirement as 'complete assignments without direction from the manager', we can easily judge if the work was finished and if it required involvement from the manager.

Once these standards are developed, the performance review system will also need to be changed (so that people aren't being evaluated on their behaviour but on their accomplishments instead). Old habits will die hard in this instance. Unconsciously, workers and managers will slip back into the habit of focusing on behaviour and activity (such as asking someone to 'make some calls' or 'do some work on that memo' or 'do some research') rather than stating an outcome or result that is measurable.

Coordination not control

Related to this shift from focusing on behaviour to focusing on performance, managers will need to give up efforts at control. Part of the reason that virtual work is initially unappealing to many managers is because they do see their management role as one of control – and control is difficult when workers are distributed by time and location. Instead, managers need to shift their perspective to one of coordination. Rather than dictating exactly what happens (and when it happens), virtual interaction empowers individual workers to take initiative and provides freedom and autonomy. While a shift to performance results in more accountability for individual workers, the nature of virtual work minimizes the ability to control them. So for managers to assume that their role is one of control is to adopt expectations that will be unfulfilled in a virtual work environment.

A good example of this is provided in the distinction between what Kimball and Mareen Duncan Fisher call the control paradigm and the commitment paradigm, as listed in Table 6.1.[3]

Whether you agree with all of this is not as important as it is to identify what we might call 'old-school management' (which is traditionally very controlling and authoritative) and 'virtual work school of management' (which is more focused on coordination, alignment and involvement). The 'virtual work school' isn't about being soft and fuzzy or not focusing on bottom-line issues and measurable results. But it involves a combination of more freedom and autonomy for staff with much more accountability and responsibility (which comes from the performance focus).

Consequently, managers who are successful at virtual work settings are not ones who manage the same way they typically would in more traditional, face-to-face work settings. Managers who function well in virtual settings are those with excellent coaching skills and good performance management ability. In other words, virtual managers need to be very good at setting goals – goals that are measurable and realistic. And they need to be good at linking people and then stepping back and letting them perform.

Table 6.1 The control and commitment paradigms

Control paradigm	Commitment paradigm
Elicits compliance	Engenders commitment
Believes supervision is necessary	Believes education is necessary
Focuses on hierarchy	Focuses on customers or clients or solution
Bias for functional organizations	Bias for cross-functional organizations
Favours audit and enforcement processes	Favours learning processes
Believes in selective information-sharing	Believes in open information-sharing
Believes bosses should make decisions	Believes the best decisions are collaborative
Emphasis on means	Emphasis on ends
Encourages hard work	Encourages balanced work
Rewards conservative improvement	Rewards continuous improvement
Manages by policy	Manages by principle
Encourages agreement	Encourages thoughtful disagreement

Source: Fisher and Fisher (2001), p. 33.

Communication as work

Unless you happen to work in a position (such as a public affairs or public relations job) where your primary role involves some form of public exposure or representation, most professionals would not view communication as their work. Interacting with others is something we regard as an enabler – it makes it possible for us to do our job, but communication is something we do in order to accomplish our work. Some may even view work-related communication as a hassle – something that keeps us from doing our real job. Virtual work environments require a change in this perspective. Virtual team members need to view communication not as an enabler but in some instances as an end in itself. Managers must regard communication as part of the job description (not only for themselves but for all their staff as well). Thus managers must not only communicate frequently with staff, they must encourage staff to interact with their co-workers as well. This will obviously create some challenges. It not only means a change in the way that some individuals work, but also challenges the perspective many professionals have of virtual work. Ironically, some people who think that virtual work is 'too isolating' may discover that the communication requirements (to frequently 'touch base' with co-workers, work to stay 'in the loop' and build virtual *esprit de corps*) of effective virtual work may leave them feeling like they haven't had enough privacy and individual time. Having talked with a number of clients who function in virtual workspace for a lengthy period of time, they mention that one of the frustrations of the set-up is that (when done well) they seem to spend more time sharing information and communicating than they do actually focusing on work. Some individuals who value solitude and prefer working alone may initially find virtual work frustrating because of the amount of interaction that needs to take place to ensure coordination and community.

It is all too common for 'silos' or 'stovepipes' to develop within virtual projects, where team members are proceeding merrily along, oblivious to what the others are doing. Especially with very effective virtual team members (where initiative is common), it is easy for everyone to charge ahead and produce work product that doesn't fit with each other. The result then is a project with major gaps. While detailed project management plans are helpful as a way to address this issue, the best answer is continual, ongoing and frequent interaction among all the team members. Because face-to-face teams can often directly observe their co-workers actions and work product, formal coordination efforts may not be as important in those settings. But in virtual work situations, it is vital that managers and team members coordinate with each other throughout the work. A rule of thumb for virtual work situations is to 'over-communicate'. Thus, managers need to regard continual communication to coordinate work as a key part of their role. Absent this perspective, it is almost certain, despite how good the project plan is or how competent the team members are, that the work will go astray and the final results will appear haphazard.

We saw in Chapter 5 (on trust) how the lack of a clear goal or the perceived absence of a confident leader creates distrust in virtual settings. While managers don't need to micro-manage and intervene constantly, it is important for them to be 'visible virtually' (through e-mail and feedback). This virtual presence (even through just short notes or quick reminders by the manager to staff) helps to diminish any perceived uncertainty within the team and build confidence in the assignment and trust in each other.

Thus managers in virtual work situations will need to view communication as part of their work (as well as part of the job assignment of all team members as well). Additionally, they'll need to think about communication as an outcome rather than an activity. Instead of 'making calls' or 'talking to staff' (which are activities), the manager needs to think in terms of 'client up-to-date on project status' and 'knowledgeable staff' and 'team member buy-in on most recent budget changes'. That means managers will need to identify what the outcome or intended result from the communication is supposed to be so they are assessing and tracking outcomes rather than activities (i.e. being a good communicator doesn't mean making a lot of phone calls and sending out a lot of e-mails). It also means that good virtual managers will need to identify potential communication breakdowns and periodically check to make sure everything is working from a communication standpoint.

Autonomy not isolation

There is a tendency for many managers to assume that since virtual teams work without face-to-face contact then they are to be left alone (especially if they appear to be successful at their virtual work). Perhaps this is due to some confusion over what face-to-face and interaction contribute to the work. Or more likely, this is an assumption that because successful virtual workers appear to be self-directed and have initiative that they should therefore be disturbed as little as possible. This thinking is nonsense. It contributes to isolated workers who feel like they are not part of the organization and are excluded from key information and events. Managers cannot treat successful virtual workers as if they are autonomous, free-floating amoeba. While virtual work may free many workers from being tethered to a geographic location or a specific work schedule, these workers still need connections to the organization and direction from their manager. The virtual nature of their work makes it even more important that their manager provide some degree of connection to the firm to help convey organizational culture and a sense of what is happening outside of their virtual cocoon.

New skills for new work

People who appear to be successful in traditional work settings are not always those who will succeed in a virtual one. Ironically, some individuals who are 'problem employees' in traditional work settings (because we focus more on their behaviour, because they prefer more autonomy in their work, because of their continual need to 'check in' and 'touch base' with others) may flourish in virtual settings. Thus it's critical not to assume that someone who is successful in a traditional setting should get priority at a virtual project (or that someone who hasn't flourished in face-to-face work is inappropriate for telework). The competencies required for successful virtual work will vary with each company (given the nature of the organization and its culture, the tasks required, the work structure and the technology involved). However, while we look at the role of manager, it is important to note that managers must be clear about how the change in the nature of the work will change the nature of the worker. Or the manager needs to identify what new skills an employee must have in order to succeed in a virtual setting. A number of books provide a generic list of 'virtual work' competencies. I caution you against accepting any of those lists verbatim. In my experience, the competencies will vary greatly from company to company.

Let's examine just a few of the variables that may be important for identifying potential competencies for a virtual worker. Obviously, the nature of the work and the structure of the job are factors. Take the example of a teleworker or telecommuter. An individual who works from home will face tremendous temptation (ranging from home chores that need doing to the ready presence of the refrigerator with plentiful snacks). Besides these temptations, home workers also don't have a whistle that blows when the shift ends. When work is just down the hallway, it is too convenient to turn on the computer to 'finish the last paragraph in that memo' after dinner is over or the kids are in bed. Consequently, many first-time teleworkers discover that they spend more time working than they did when they reported to the office outside their home. Therefore one important set of competencies for teleworkers involves discipline around work – both the ability to prepare and start on their own, but also the ability to quit – to stop work. This last point is important to consider. If the teleworker isn't good at stopping, firms will quickly find that staff working in such situations face failing marriages or home life, their social life disappears and resentment to the firm grows (because the teleworker quickly realizes that they are putting in longer hours than their counterparts at the main office yet probably not getting compensated for it). Work/life balance is important and teleworkers that don't know where to draw the line quickly find their balance disappears. Virtual team members require a great deal of initiative. If you recall our discussion in an earlier chapter on trust, initiative by team members plays a significant factor in the development of trust and cohesion. For an individual worker (such as a teleworker or telecommuter), this is less of a factor (since they are likely to have more direction) than it is for a virtual team. And as mentioned earlier, there needs to be a strong interest and comfort in information exchange and communication for most virtual workers. Thus a wise manager will keep these factors in mind as they hire or select personnel for virtual work.

Ironically, while many people consider virtual work to be 'antisocial' and believe it encourages isolation, a key competency group for all virtual workers involves communication skills. The range of communication skills varies with the technologies involved and the nature of organization and type of work. But communication skills are probably more important for virtual workers than they are for face-to-face workers.

For virtual work that relies a lot on threaded discussions, document exchange and e-mail, writing ability is vital. This sounds like a statement of the obvious. However, in a traditional work setting, managers can often call on others to generate reports. White-collar workers can use spell-checkers or grammar software to enhance their writing. But in virtual environments (where e-mail exchanges are often fast and furious – with some workers dealing with 200 or more email messages in a typical a day), the ability to review, rewrite and spell-check all messages is impossible. Such environments call for the ability to 'write out loud' (much like giving a speech extemporaneously), organizing text quickly and writing coherently. Additionally, some research with virtual teams seems to show that spelling and writing ability significantly affects trust within the team (individuals that write poorly inadvertently feed a perception that the team may be struggling or that as individuals they are not credible).

Additionally, good virtual workers are people who deal well with change and fluidity. They are individuals who are capable of creating order out of disorder – or coping with uncertainty. Because virtual settings often arise out of necessity (rather than out of planning), they are often circumstances that involve little infrastructure, short timeframes, projects that are not well thought-out and crises requiring quick action. Even when a virtual project is well thought-out or planned, it is often done virtually in order to allow for more flexibility. Thus professionals who flourish in virtual work settings are good at sorting out what needs to be organized or structured and what elements can be left as they are (fluid or chaotic). Individuals who have a strong need for structure and don't deal well with uncertainty don't handle virtual work well.

Because virtual work involves new skills (or places greater emphasis on existing competencies), there are several implications for managers. Managers must be good at sorting out who will work well virtually and who won't. That means a good virtual manager will have identified what skillsets will be more important for the new work situation. Managers must also be good at coaching and developing these skills within their employees. Ideally, a manager would be able to select membership for a virtual team from a cast of well-qualified members. In reality, the manager may be forced to choose some team members because of their subject-matter expertise – despite their obvious deficiencies in virtual work competencies. In those situations, the manager must work to develop those skills within their employees. This may be through one-on-one coaching (in-person or virtually) or it may be through more formalized training. In any case, a successful manager will not assume that someone successful in a traditional work setting will automatically succeed doing virtual work.

Let's look at an example to see some of these principles in action.

> Isabelle is the senior manager for the IT department of a company with offices throughout France. Headquartered in Paris, most of her staff are located there. However, there are field offices in Nice, Lyons, Nantes, Marseilles, and Toulouse in addition to the Paris location. All told, she manages 14 staff members (five at field offices and the rest in Paris). She occasionally makes site visits to the other locations. There is only one IT staffer at each field office. The IT staffer for that office handles a combination of installation duties (software upgrades, hardware set-up) and troubleshooting (dealing with computer problems on-site, collecting local information (such as sales data) and sending it to Paris (to be compiled and placed on the company intranet). Because the intranet must be kept up 24 hours a day, seven days a week, several of her staff are on shifts – working nights and/or weekends (meaning that they are not usually there when Isabelle is). Consequently, Isabelle manages a team that is dispersed by time and location.

Isabelle finds managing her team to be an extremely frustrating experience. The staff in Paris during the weekday shift (when Isabelle is there) have the most camaraderie with Isabelle. This is evident in conference calls where the staff in Paris joke with her and refer to what is happening at headquarters while those calling in (either from the field or at home if they work evening or weekend shifts) seem less 'connected'. Isabelle complains that she has difficulty managing the field staff because she can't see what they're doing. Consequently, she demands a lot of reports from them on their activities. At one point, she even insisted that they keep time logs (indicating how much time they spent on particular activities during what part of the day). She will sometimes come in early or stay late in order to watch the shift workers, but is worried that after she leaves they may 'play around'. Isabelle has always placed a lot of emphasis on how important it is for her staff to be busy. Now, with a higher turnover rate in the field offices (where she is having trouble keeping people), she stresses the need for staff to have an even higher energy level. The staff members complain that they are being run ragged. Staff members (especially out in the field offices) complain of not being 'in the loop' and being uninformed – something Isabelle writes off to the nature of being away from headquarters.

What should Isabelle do differently? What advice would you give Isabelle?

For starters, Isabelle needs to shift from managing activity and behaviour to a focus on results. Given the location, time schedules and number of staff, there is no way she can observe even a fraction of the work and still handle her other responsibilities. This particular organizational set-up (with field offices and shift work) is ideal for performance-focused assessment. Since outcomes or results exist after someone leaves work, she can assess what people are doing and how they're doing it by choosing the correct results to focus on. This is particularly true of IT, where she should easily be able to identify measure like: lines of code generated or debugged, percentage of time the system was down, online sales totals through the website, response time to staff requests for troubleshooting. In fact, many of these outcomes would have financially quantifiable results (decreased cycle-time saves money for instance). So Isabelle needs to shift from focusing on behaviour to focusing on results.

Her efforts to measure behaviour (the time logs for instance) not only are failures, but in this instance probably breed distrust. She doesn't appear to be doing anything to provide any sense of community or cohesion among the field offices or shift personnel – so they feel like outsiders. Combined with the apparent distrust, it's no wonder there is a turnover problem.

Nor is she pushing communication between her staff. While interaction does take place (such as at staff meetings through conference calls and discussions about work product), it isn't clear that she encourages it. Nor has she made the simple effort of assigning mentors or set up a 'buddy system' so that field staff have connections with other personnel (to check in with each other and provide some linkage to the rest of the organization besides conference calls and e-mail). Without any of this, field and shift personnel probably feel isolated – as if they are a department of 'one' rather than part of a much bigger operation. The lack of communication probably feeds the distrust as well. In short, Isabelle is doing almost everything wrong that she could do in this situation!

Questions to consider

1. Think of a manager you know of (either in a current or past position) that you admired. What was it about them that you admired? In what ways do you think they would need to change to be a successful manager of a virtual team?

2. Think of your current job. Are you evaluated on the basis of your behaviour or your results? If you're judged on the basis of your behaviour and actions, what would a list of results or outcomes for your job look like?

3. What do you think would be the most challenging aspect of managing virtual workers?

Notes

1. This is from a 2002 study on e-mail usage at work by Experian in the UK.

2. For a more in-depth explanation of the need to focus on outcomes rather than behaviour, look at the work of Tom Gilbert who has written eloquently on the problem of

focusing on activity rather than results (see, for example, Gilbert, 1996).

3. See *The Distance Manager: A Hands-on Guide to Managing Off-site Employees and Virtual Teams*, by Kimball Fisher and Mareen Duncan Fisher (2001), p. 33. Trina Hoefling (2001) also discusses these managerial issues in some depth.

CHAPTER 7

New work order

'I get less done when I'm at headquarters than when I'm working from my home office. At headquarters, I feel compelled to check in with everyone. I face a lot more interruptions. I'm just not as productive as I am at home.'

Morgan Ford, teleworker, Marasco Newton Group

Virtual teaming is a form of telework. But telework involves much more than virtual teams – dispersed teams are just one subset of telework. If we're to understand virtual teams, it's important to look at telework itself. This is because many virtual teams first start with one or a handful of individuals working from home or working on the road. Then, as the organization begins to rely more on teams, thus teleworkers become part of virtual teams.

Today's workforce has seen an explosion in the amount of telework and virtual arrangements within organizations. According to one recent study on virtual work in the United States, over 46 per cent of all American workers are virtual to some extent. In companies of over five hundred employees, the numbers are even higher – 61 per cent have some kind of virtual work arrangement.[1] Although the United States and Singapore are the world leaders in virtual work arrangements, the rest of the world is quickly adopting virtual work arrangements as well. The EU has seen an increasing reliance on virtual work. 'Telework is a phenomenon that is nowadays known in all parts of the European Union' (Dangelmaier et al., 1999).[2] The projected increase in telework in Europe alone is startling. 'The numbers of people working from home or on the move could reach over 27 million by 2010' according to the latest EMERGENCE study, *Modelling eWork in Europe* (2002).[3] The Gartner Group estimated that more than 137 million workers worldwide will be involved in some kind of remote work or virtual work arrangement by 2003.[4] Even if these numbers are wrong, there are at least two clear conclusions we can easily draw. First, there are a lot of organizations in a lot of countries that are either expecting or allowing workers to participate in some form of virtual work arrangement. Second, most organizations are not planning for or preparing adequately for telework. In other words, while many organizations are either engaging in telework or are going to be doing so in the near future, very few of these organizations (as well as those companies that resist virtual work arrangements) have approached the issues around telework with an informed and intelligent analysis. Thus we see decisions to allow or discourage telework in organizations for all the wrong reasons (or no reasons at all – the virtual work arrangement accidentally evolves).

What is telework?

Before we proceed, it makes sense to clarify some terms. Telework and telecommuting are often used interchangeably. While there is a tremendous amount of overlap between these two terms, they actually refer to slightly different virtual work settings. Telecommuting is a term that is generally agreed as having originated in the United States. While the term telecommuting is often used by some to refer to all forms of virtual work arrangements, it first came to mean (and most accurately applies to) situations where an employee who reports to a regular office job is also sometimes allowed to work from home. Thus a telecommuter still typically reports to the office of an organization in which many others may also work – but may only report to that office for 50–80 per cent of their typical work week. Again, while telecommuting in the United States has come to mean much more than this, the concept was initially viewed as a means to reduce the costs (gasoline, pollution and commuting time) of employees and organizations by allowing staff to do some work from home. However, the term still implies that the worker is tethered to an organization hub (a formal corporate building) rather than the idea that one's true office can be at home or on the road or in the field.

Another term that is not technically part of virtual work but ends up having applications for telecommuting is that of flextime. Flextime is a concept that is now widespread in a number of organizations (such as the US government). A flextime work arrangement means someone is working in a central office that they report to, but they have flexible work hours as to when they arrive or leave. In a flextime arrangement, it is not atypical to have periods where some staff report as early as 6 a.m. and others start their day as late as 11 a.m. This does not even consider other flextime arrangements (such as compressed work weeks, working weekends or shift work). However, it is obvious that flextime arrangements create situations where not all of the staff are at work at the same time. Even with designated 'core times' (where staff are supposed to be at work – often this is the 11 a.m.–2 p.m. period), because of health issues, client commitments, commuting problems and the like, the 'core time' concept doesn't ensure that all staff are available for key meetings on-site. Consequently, though flextime isn't supposed to be a virtual work arrangement, any organization implementing flextime needs to treat it as if it were a virtual work set-up. This is because, inevitably, staff will contact their peers at home to answer key questions (because 'only Nigel knows the answer and he hasn't come in yet!'). So issues about when is someone 'off the clock' and how to work 'virtually' become critical for effective flextime situations.

Telework is a term that is much more widely used in Europe than the United States. Again, telework has often been treated as a synonym for telecommuting. While telework is sometimes used to refer to staff who are allowed to do some work at home (and still do some or much of their work at a corporate office that other staff report to), typically this term has a broader meaning than that of telecommuting.

The next step (if we were to view varieties of virtual work on a continuum starting with the most traditional set-ups such as flextime and telecommuting and moving to less traditional arrangements) would be that of office-tethered workers. In an office-tethered arrangement, a worker still has an office they report to but they travel extensively and often do most of their work out in the field. Many sales representatives (who spend the bulk of their time visiting clients outside the office) fall into this category. They may attend a few meetings in the office, show up once a week to fill out travel records and sales reports but spend most of their time on the road. Road crews (working on repairs), utility repair staff (fixing downed lines or installing telephone/cable for homes) or police officers are all examples of office-tethered professionals. They face a range of virtual work challenges (how to stay 'in the loop' when they are out of the office, how to collaborate effectively with people they don't see most of the time, how to manage people who are out of sight), yet they usually aren't viewed as virtual workers. Instead, they are typically seen as being assigned to a specific office or location, even though most of their time and results are generated outside of that office.

Many office-tethered workers enjoy this role because they feel they have more freedom than those peers who stay in the office full time. If so, this is probably because the organization still operates from a 'control paradigm' that focuses more on what people do or how they behave instead of on what they produce or the outcomes of their work. Ironically, many professionals who insist they couldn't function in a virtual work arrangement flourish in an office-tethered set-up because they don't recognize it for what it is – a virtual workspace. In part this is because they associate 'virtual work' with isolation and technology. If someone is a 'technophobe' or likes being around people, they may initially be very resistant to the concept of telework – because they associate it with separation from others and connecting only through technology. As we've seen (and the office-tethered case is just one example), neither of these charges (relying on technology and social isolation) is necessarily true of virtual work situations. This is an indication of how important perception and mindset is to success in a virtual work setting.

Hotelling refers to temporary office arrangements or 'hot desking'. The term 'hot desking' is taken from the US Navy's concept of 'hot bunking' where several sailors (who all work on different shifts) share a single bunk on a submarine. There may be two or three sailors, but only one of them is sleeping at any given moment while the others are on duty, so they all share the same bunk. Typically, a hotelling situation is one in which an executive or professional travels to a site and works there for a period of time (such as several weeks) and then moves on. That professional is assigned a desk/workstation/resources that is theirs for the limited time they're at that site. When they move on (and another professional moves in), that person takes over those resources. Almost all large organizations have unassigned work space that is used to house temporary workers or staff that 'drop in' for periods of time. In some cases, this 'work space' may be nothing more than an empty desk and chair. In other organizations (with key personnel who regularly visit worksites), the temporary resources may include IT and secretarial staff dedicated to the hotelling professional, special file sharing protocols, security arrangements and so on. Today, hotelling is a common organizational response for the proliferation of consultants working on-site with long-term projects. For instance, I once did some consulting work at the US Embassy in Moscow. While some of my work could occur virtually (away from the Embassy), much of the work required face-to-face interaction within Moscow or access to resources in the Embassy. During my approximately four weeks of work at the Embassy, I had a temporary desk, computer, limited access to IT capability and when I was gone the 'desk' was taken over by someone else.

Hotelling, though common (both out of necessity and as a cost-savings measure) is not a popular strategy for most participants. Many individuals who participate in hotelling (especially internal staff – people who are employed by the company they hotel at) tend to feel like second-class citizens. File-sharing protocols and document access also become major issues in this work set-up.

From hotelling we move to the home-office situation. In a home-office set-up, a worker operates entirely from home. They are tethered to an office, but it is an office they don't share with anyone else (at least from the same organization). In the home-office set-up, it is very easy for an employee to feel disconnected or an outsider from the organization they work for. Even though a home-office worker may not undertake any travelling (unlike the office-tethered staff member or the hotelling professional), these other groups at least feel some degree of physical connection to a shared worksite with others. At some point, they return to headquarters or their travel ends. If the home-office worker is at headquarters, it is for a 'visit' or to attend a meeting. There is no confusion over whether they are based at that site.

The home-office arrangement is frequently the source of many misperceptions and misunderstandings. Organizations typically believe (falsely as it turns out) that workers in such set-ups will spend most of their time watching television or goofing off. Professionals who think about working from home tend to see it as a panacea – where they'll get everything done but skip the commute or be able to do their work near the neighbourhood pool or outdoors on the porch. The reality usually ends up very different from both perspectives. Home-office workers typically have difficulty setting boundaries and tend to work longer hours than they would if they were reporting to a corporate office. It is just too tempting to go back to the computer after dinner or once the children are in bed. Organizations typically get more work than they deserve from home-office workers. Professionals contemplating a home office fail to realize the different kinds of interruptions they face (such as the neighbour knocking on the door because their nanny is sick and they need someone to watch their child while they go off to their corporate office or the door-to-door salesperson). Home-office set-ups can also be very isolating for many professionals. Consequently, the home-office work arrangement is one that has the greatest number of initial misperceptions.

On our virtual work arrangement continuum, the last (and most virtual) of the set-ups is that of the fully mobile worker. This is a professional who is not only not tethered to a company office (permanent or temporary office), they aren't even tethered to a home office. They work almost entirely in the field. Such professionals would typically admit to having two offices – a 'formal office' that exists on paper and their 'real office' which goes with them as they travel. The 'real office' is often a laptop and wireless phone. They may have a home office or a company office but they travel almost 100 per cent of the time and their formal office tends to exist so they can collect mail and have an address on their business card. Unlike office-tethered professionals, they don't return frequently to their formal office and it is not regarded as a base of operations. Obviously, those who perform in these circumstances are a truly unique breed of people and this kind of work set-up is not for everyone.

As you reviewed these telework arrangements, you probably felt there was some overlap – or instances where some work set-ups fell into two categories. For instance, some telecommuters may also have a home office. Or, during a particularly busy season that requires many field visits, a flextime work arrangement may also include a heavy dose of office-tethered set-up (because of the need to travel). The value of these work set-up distinctions is not that each virtual worker cleanly falls into one distinct category. Rather, each work set-up tends to produce particular dynamics. Someone who works only from a home office faces different challenges to someone who has a home office but still does most of their work from a corporate location. The value of the work arrangement distinctions is to help understanding of how each of these set-ups requires different things of workers and their organizations.

How not to telework

Much like an organization's experiences with virtual teams, telework and virtual work situations duplicate many of the same mistakes and mis-steps. The two single biggest mistakes about telework that organizations make (among those who do have virtual work set-ups) are as follows:

- They don't consciously choose particular telework arrangements but accidentally back into them. Instead of planning for telework and attempting to direct the path it evolves towards within the business, virtual work just 'happens'. The executives and staff wake up one morning to discover that crucial functions are being done virtually. Sometimes, the virtual processes that have evolved are very functional and efficient. In many cases they run counter to other key organizational imperatives or opportunities to leverage the virtual work with other actions have been lost. For instance, out of necessity (because the sales force spends so much time in the field), each of the sales reps has acquired their own wireless telephone. But because they've done this on their own (rather than the company acquiring the phones), each rep has a different phone system, different number assignments (so they aren't numbered sequentially), and each phone has different capabilities. Conference calls, shared text messaging, even accessing voice mail at their home office – all these opportunities have been lost because no one anticipated the situation and looked for ways the company could maximize the value from the virtual interaction. A comparable circumstance would be one in which a company has each employee using computers and software that are compatible with all others in the company – as opposed to one in which each staff member acquires their own computer and loads software (and operating systems) of their own choosing. So Mary has an Apple machine, Trevor is using Linux on his PC, Gina has Windows and there are 17 different word processing packages (and 5 versions of Word) on the various computers in the company. That would be a situation of utter chaos with almost no interoperability.

- Telework arrangements are adopted for the wrong reasons. Many of those pushing for various forms of telework are employees. Staff members think they would like to work from home – so management grants them their request (with the belief that a happy worker is a good worker – or more specifically, if management grants them this request to have a home office or allow flextime then they'll be able to get the workers to tolerate or accept some management demand on another issue). There may be times when telework as a quality of worklife issue is necessary for organizational competitiveness (for instance, two companies compete for the same workforce and the first company has a reputation as a much better place to work). However, in general, I find that telework initiatives that are adopted either because of an issue of employee morale or as some part of a quality of worklife effort are usually doomed to fail.

This does not mean that organizations should not be concerned with employee happiness or quality of worklife. However, what is almost always true is this: if a programme or initiative is not strategically important to an organization, at some point it will no longer be a priority. For instance, if an organization agrees to adopt a telecommuting programme because staff hate the daily commute into London, that telecommuting initiative will be tolerated only as long as it doesn't create a challenge for a business imperative. As soon as a major business crunch occurs (such as a key proposal that requires work or a heavy stretch of work such as a busy season for accountants), management calls a halt to the telework initiative.

Telework initiatives that succeed are those that are driven by major business goals or have some strategic driver within the organization. For instance, sharing knowledge is generally a good thing – who doesn't want an organization with lots of informed staff and lots of knowledge? No one certainly wants the converse (dumb, uninformed workers). Yet, organizations are willing to tolerate spending on 'knowledge management' and the systems that are part of it only as long as times are flush and it is convenient to do so. At some point, the budget gets tighter and management asks: 'What is the return we're getting on all this knowledge management spending? What money has it brought in? Which of our strategic goals are we closer to meeting because of it?' This applies equally well to telework initiatives. It is all fine and good to adopt telework programmes to help working mothers or convalescing employees and allow greater staff freedom so they can work from home. But these initiatives won't stick with most organizations unless there is a tie-in to a major business goal. For instance, if allowing home offices has a measurable impact on customer service (because using home workers allows for longer periods of phone coverage – because they have a two-minute commute from the bedroom to their office, not a 40-minute commute to the corporate office) and if that impact on customer service is measurable to some degree, then as long as customer service is valuable to the organization then the home-office initiative will have strategic value to the company.

It is not enough to demonstrate that telework options will save the company money. There is very strong evidence from a range of sources that various telework arrangements can decrease the cost of office space, save on energy bills for the company and decrease costs for employees (such as dry cleaning and commuting costs). But for most companies, these costs are not assigned to a specific line item and it is often difficult to demonstrate for a specific company how a particular programme (such as flextime) has resulted in savings. What has far more impact is to identify a strategic goal within the company and identify how telework supports that goal in a key manner.

While there are many other factors in the success and failure of telework programmes in organizations, these two points (no intentional programme and doing it for the wrong reasons) are the two biggest reasons for failure. Organizations seeking to do telework initiatives typically want to see models from other organizations that have 'done it right'. What they fail to understand is that successful telework has less to do with how it is structured and far more to do with motivation (why are we doing this?) and consciousness of effort (choosing to operate virtually rather than backing into it). All organizations differ in many degrees. Thus identifying a structure from a company that has enjoyed success is not particularly useful as a tactic – because the culture and assumptions will vary with each company (and thus what works for company A won't necessarily work for company B).

Factors to consider

The complete list of factors to consider regarding whether or not to adopt a telework programme will vary for each organization. For instance, a unionized workforce faces different telework considerations than a non-unionized one. The nature of existing technology raises considerations. How geographically dispersed the organization is (will there be a need to coordinate across time zones? across different languages and countries?) raises considerations. However, there are several factors that every organization considering telework initiatives needs to consider.

- *Which model are we considering adopting: departmental, corporate, individual or contractor?* This question considers how widespread the telework initiative will be and what unit of work will participate. The contractor model involves a number of different formats. It could consist of opening up telework options to those who collaborate with the company but are not officially employees (consultants or partners on projects). This model may herald a company transition to greater use of consultants and 'outsiders'. The individual model is one in which the company is determined to keep a few particular individuals (perhaps a programmer going on maternity leave who is vital to a major project or a webmaster who is important to the company but is only consistently needed twenty hours a week). The corporate model means that everyone in the company is going to do it – a full immersion strategy (although this may be phased in over time either by work unit or the type of virtual interaction, for example as beginning with telecommuting then moving to hotelling). The departmental model is one in which a unit within the organization (such as sales or IT) becomes virtual while the rest of the business remains face-to-face. There are arguments for each model. Which one is best for your organization? The answer is that which model is right depends upon the specifics of each individual organization as well as their motivation for doing the telework initiative.

- *What are the expectations of all of the relevant players?* Most managers and staff have wildly inaccurate perceptions of what a specific virtual work situation will be like. Additionally, most players are clueless about what some of the key issues involving virtual work initiatives will be. For instance, most staff arguing for home-office options probably haven't thought through the issue of how to draw the line at work – where and when to stop working. This seems like a ludicrous concern to most professionals – how to stop work! For most people who report to the office, they can leave the office and that tends to establish some kind of boundary with work. But a home office eliminates most natural work–home life boundaries. Most teleworkers have discovered that with a blurred line between home and work, they tend to work longer hours.[4] Thus any organization contemplating telework initiatives must find a way to expose and examine assumptions and expectations by everyone, but also needs to identify what assumptions the participants are unconscious about. What assumptions are they making that they aren't aware they are making? What expectations should people have about this initiative? A good telework initiative confronts those questions and puts them on the table for people to examine. Even if you feel like you've done no more than partially answer these questions, by exposing them you get people to recognize that telework is a brave new world with lots of unexplored territory. When unexpected surprises then occur, people are less likely to be upset or feel betrayed. Instead, they recognize that something that goes with the territory of exploring something new is that they're going to encounter a surprise now and then.

- *How will we build in interaction and interpersonal support?* Even introverts need some human support at times. It's a mistake to think that home-office workers have no human contact – they are likely to have plenty. But one of the problems with the nature of telework is that the contact with the rest of the firm is likely to be mostly task-oriented ('James, when can you have that report done?' 'Larissa, I need your response to the report by close of business today!'). Some of the visual cues that co-workers might notice when someone is struggling or emotionally troubled are usually missing in telework situations. Consequently, a home-office worker might be participating in several hours of phone conversations or conference calls each day, answering a multitude of e-mail messages and tapping into videoconferencing sessions, yet feel incredibly isolated. What has happened is that all of the interaction and stimulation they're likely to receive in the course of a normal day (to the extent any business has 'normal' days) is focused on immediate task priorities. So issues that mentors, friends and support networks might deal with in most typical offices end up being avoided in home-office situations. This is not to say that telework set-ups can't deal with these situations, only that this kind of interaction tends to be absent (because we typically rely on non-verbal cues to tell us that Paul or Gina has a personal issue bothering them).

Thus it's a good idea for organizations with home-office staff to either encourage staff to devote a certain amount of work time to 'housekeeping' and social exchange or to create a 'buddy system' where co-workers are assigned to others with the job of checking in with them to see if they have concerns or issues.

- *What kind of bonds do we want to create or remove?* Telework situations tend to have significant impact on a number of bonding dynamics within an organization. Because a home-office worker spends less face-to-face time with peers and may rarely show up at the corporate office, loyalty to the firm often diminishes substantially. This is an important point to make – too many firms adopt telework policies with the belief that 'if we do this for them then they'll owe us' or 'they'll be more loyal because we've done this special favour for them'. The reverse tends to be true. It's not that the teleworkers resent the new work situation or respect their employer any less. But the physical elements that tend to subconsciously create an identity and affiliation with the company (such as the headquarters building and the presence of all their friends) are now all gone. The teleworker discovers that they don't need to be in the building to have contact with all their peers or friends and that leads to the discovery that they probably don't even need to work with the company to stay in touch with these people. Someone working out of their office starts to realize that regardless of their affection or thanks to their employer, they can work for lots of different organizations if they really wanted to. So, the idea that telework policies are automatic answers for retention problems is a wrong one. Telework and home-office policies can improve retention, but not through loyalty. Instead, a home-office set-up can eliminate some of the considerations that force a valued employee to choose between staying versus quitting (or moving to a competitor). So, telework arrangements can contribute to retention. But they don't typically do so through enhancing loyalty because the company has provided a special service or favour to the employee. They simply remove some of the reasons why the employee was considering leaving in the first place. Related to this, it is natural for many teleworkers to feel like they are no longer 'part of the team'. Now remember our discussion on virtual teams and trust from earlier chapters – of course teleworkers can be in groups with remarkably high degrees of trust and cohesion. However, many home-office set-ups are where an existing workgroup sees one member (who is thinking of retiring, is hurt, goes on maternity leave or is tired of the commute) who goes on a telework arrangement while the others continue to work in the corporate office. Absent efforts to build trust and cohesion, that new teleworker is likely to feel like an outsider. This issue may not be especially important to the company. But if the work processes require a great deal of collaboration and teamwork, then companies will need to address this issue if home-office set-ups are not going to be a negative factor in performance.

- *How will we inculcate organizational culture and values?* There are a number of very successful organizations that recognize the importance of a company culture and set of values that are embedded in the workplace. Companies like Disney or Royal Dutch/Shell or Hewlett-Packard all have strong pervasive cultures that make them unique among many of their competitors. The culture each of them possesses provides the company with a competitive advantage, makes for a more cohesive workforce and helps provide guidance to people (by implying how things are done, the way people are to be treated and why the company is so special). Teleworkers tend to be more removed from the presence and influence of the culture. We see this anyway with field offices (which often have remarkably different cultures from the rest of the parent organization because of their geographic separation from the other staff). However, when workers are isolated further (by working in home offices), they are even further removed from the presence and influence of the corporate culture. In some cases this is a good thing. A business with a dysfunctional and sick culture (such as one that penalizes innovation and initiative) will have less influence on home-office workers than it will on those in the corporate office. But the reverse is also true. Organizations that gain significant value from their culture will discover that teleworkers are less influenced by it and more removed. Thus if organizational culture is a significant asset for a company, companies need to carefully consider home-office set-ups. The company will need to build in special ways of stimulating and maintaining company culture and values within the teleworkers.

Logistics

Working from home requires a surprising large amount of forethought and planning about set-up and equipment details. Things that we take for granted while working for someone else at a corporate office suddenly become important concerns once we're operating out of the home. Here is just a short list of some physical and logistical considerations for any home-office set-up.

- *A good, comfortable chair.* Because you won't have as many interruptions or office meetings at home, you will probably spend far more time sitting in a home office than you would in a corporate office. A good home office must have a chair that is not only comfortable but sturdy and ergonomically superior. From personal experience I can't emphasize how important this aspect is. If you expect to spend six to nine hours per workday in your office, you're likely to spend the bulk of that time seated. So get something you can sit in that won't wreck your back and distract you from your work. Remember, too, that most office chairs will tear up your wood flooring or carpet or scar your floor tiles, so you're going to need some kind of floor protection to go with it (such as a sturdy floor mat).

- *Furniture.* Related to the issue of the chair is that of furniture. You probably don't need the same resources that people in the corporate office do (because there are issues of equity and uniformity that at the corporate office become less important when you're working at home). But depending upon the nature of your work, you may require a workstation, file drawers, printer table, conference table and shelves. There is also a host of smaller 'equipment' needs that may be relevant: anti-glare screens (because home lighting is less likely to be fluorescent), reading holders (to free up hands to type on a keyboard instead of keeping a book or report open) and lighting. It's not that all of this equipment is essential, only that the equipment needs of corporate workers and the equipment needs of a teleworker may be very different.

- *Access to information* (or your own personal professional library). Maybe you'll be able to access company resources online or over the phone. Or maybe you're lucky enough to have a public or university library close enough that you can easily use to get information on software programs or to research client information. But more than likely you're going to start developing your own home-office library. It will probably need to include manuals and support materials for any technology or software that you use. You'll need copies of any style sheets and memo guidelines (for the documents you produce for the company that you won't have nearby copies to compare with). You'll need to set up files for the work you do (and all the company information: memos, e-mail, reports) that is on a bulletin board at work or taped to the manager's door (that you took for granted because someone else made it available) that now you either need to file or lose access to it. And while we're talking about access, most large companies have firewalls that limit the ability of outsiders to tap into their phone or computer systems. Teleworkers will require gates through this security.

- *Technical support.* A good teleworker starts to become a 'jack of all trades' who is capable of solving many of their own problems. When the computer locks up, you can't call the help desk and have a technician walk down the hallway to troubleshoot things. Instead, savvy teleworkers become more self-reliant and learn to cope with many of the problems their corporate peers count on others to solve. But set-up of the home office may involve some additional technical problems such as the wiring in the house (installing a home office may exceed the capacity for the room in the house that the office is in) or access to phone lines. None of these problems are insurmountable, but they often need to be resolved one by one. This is because Helen lives in an old cottage with outdated wiring while Rahim is in a more modern flat. A home office for each of them will require different levels of work and different obstacles. Thus company telework logistics may often need to be resolved individually.

- *Privacy.* Depending upon the living circumstances, there may be tremendous distractions for the home-office worker. A teleworker who lives alone in a free-standing house in the country is likely to have a great deal of privacy. Someone with family or room-mates, who lives in a noisy neighbourhood, or who works in a room that opens to other parts of the house is likely to have privacy problems. In my home office, I have to beware a neighbour who cuts the lawn during the day every Friday (and the noise of the mower makes phone conversation difficult). On days when my son is home from school, we've had to adopt a family policy that says no one enters my office when the door is shut or I'm on the phone. I almost always don't pick up calls to our home number while I'm in my office. You will also need to consider privacy and security issues when office hours are over. (Young children may find your office supplies to be irresistible toys and will want to play with the copier despite your orders to stay out of the office when you're out!)

- *Basic office equipment and tools.* A home office can quickly become an expensive proposition for a company if they insist upon providing versions of all the equipment that peers have at corporate premises, such as a copier, printer, fax, phone and personal computer (let alone basic office supplies). Fortunately, there are a number of machines that combine functions. But there are also other pieces of equipment that become important to many teleworkers. Headsets are invaluable for many teleworkers (so you can type or search for files while taking information from a customer over the phone). Surge protectors fall into the category of 'no-brainers' for home-office workers. Depending upon how vulnerable the work is you're doing, it's possible you might need a back-up power supply. Another issue is what happens if the home-office worker has a fax machine. Many business faxes queue up their work and will send faxes repeatedly (or try back with numbers that were busier earlier in the day). Consequently, it is not unusual for anyone with a fax at home that has a widely disseminated number (in other words, the fax number is on a business card) to get a ringing telephone (signalling an incoming fax message) at 2 a.m. So home workers may want to look at a variety of arrangements such as fax-modems on a computer, silent rings for fax machines, services like e-fax that allow facsimile messages to go to an online site you then receive e-mail from. Teleworkers may also want to consider whether or not they need caller ID for their phones. Corporate offices (with a telephone tree) can screen many calls – home workers cannot. Consequently, there are a number of equipment decisions to consider. What a number of companies have taken to doing is to give teleworkers an allowance (which they may spend on equipment needs). Any money left over is kept by the worker, any costs above the allowance come out of their own pocket (so the teleworker has an incentive to economize and buy wisely only the equipment that is needed). This means that if you're exploring a pilot telework project with your company and they are considering the allowance approach, do some serious price shopping before you agree on an allowance figure. Otherwise, you're likely to get stuck with the initial amount (which may be wildly unrealistic and completely inadequate).

- *Business cards.* Yes, have them, but whose address goes on the business card? If it's the teleworker's home-office address, that creates potential confusion about corporate identity and location. If it's the corporate home address, have provisions been made to get any relevant mail to the teleworker? As simple as this seems, mailrooms often aren't 'in the loop' and when they get mail for someone no longer working in the building the tendency is to assume that person has left the company (in terms of employment, not location!). Even the forwarding of mail isn't such an easy solution – sending mail to the corporation and then the home office adds additional days to the process, something some companies may wish to avoid.

- *Meeting space that is convenient.* Whether or not the teleworker meets with co-workers or clients on a regular basis will depend to a great extent upon the nature of their job. However, frequently a teleworker will need access to meeting space (and there isn't any more room appropriate in the house) and driving to corporate headquarters may not be an option. Again, there are a range of options available for teleworkers (such as office suites that rent out meeting space or even meeting in a coffee shop or restaurant). The key lesson here is that teleworkers can't assume that working at home means there is no need for meeting space.

- *Insurance.* The company will need to consider what kind of insurance protection they want to provide for their employees working from home or if there is a need for any additional insurance coverage. If the employee is using company equipment, the corporate insurance policy may not cover computers and printers located at home. Records and files may also involve material the company wants insured. The kind of work the teleworker does will provide some guidance about the kinds of insurance coverage that may be necessary. Teleworkers will also want to see how their personal insurance is affected by the presence of a home office.

There are plenty of other items that might be part of a home office – this is intended to be only a basic starter list of considerations. In my experience, organizations and staff just starting out with telework pilot programmes tend to fall into two extremes when it comes to logistics. Either the teleworker (for equity reasons) is expected to have all the equipment that corporate staff have (which is almost always needlessly expensive and the teleworker eventually regrets the space lost in order to have that full-size laser printer or company-standard workstation). Or the worker and company give little thought to what kind of logistical issues exist and months into the pilot programme are forced to renegotiate the ground rules. This is because the teleworker started out using their own personal computer and when it crashed was unable to perform at the necessary levels. In any case, before you start a home-office arrangement, give some serious thought to what your office set-up needs to look like. You probably won't need much of the equipment you initially want but probably require some items (this is especially true of issues involving telephone lines, circuits and electrical capacity) that you didn't initially think of.

What to do?

Imagine that you've just recently been given permission by the company you work for to have a full-time home office. You're now going to be commuting to the spare bedroom each day instead of battling traffic. But the first month of work in the new arrangement has been baffling for you – it wasn't at all like you expected it to be. Your manager (still located at the corporate office) calls you at a range of odd hours to 'check up on you' and demands weekly reports with timesheets (in 15-minute increments) accounting for what you're doing each workday. Plus you find that now you're home, your spouse (who still commutes to a job) has taken to calling you at home during 'work hours' to ask you to begin preparing dinner or to put something in the wash she'll need the next day. She's also gotten in the habit of scheduling appointments for repair staff to show up while you're in your home office since 'someone will now be home to let them in'. In short, you find yourself with a lot more 'busy work' than you had previously (and you had a lot when you were commuting). And with the close proximity of the refrigerator, you've also started to put on a lot of weight very quickly. What should you do about this situation?

There are a couple of important lessons from this. A home office may be at home but it's imperative to treat that work setting like the job it is. That means you need to develop good work habits and insulation from the rest of the house. Neighbours may start getting in the habit of dropping by to 'borrow some sugar' or with some other request because you're home. You'll need to set some boundaries. This may mean that you refuse to answer the doorbell during your work hours. Also, have a separate phone line for work – and then refuse to answer the home telephone when you're home. These boundaries will also include telling your neighbours or spouse or roommates that you are at work – and that you're perfectly willing to do personal stuff (baby-sit a child when the parent runs an errand or get the laundry started) – but only after *your* workday is finished!

As for your supervisor, it sounds as if she is focusing more on what you do (your activities) than how you perform (your results). It's best to sit down with your supervisor and have a discussion about expectations for this work set-up. Then talk about how your work will be measured. If your manager can't come up with some measurable outcomes, you'll need to suggest some ideas of your own. Otherwise you'll be henpecked to death with requests to indicate how you spent the last 15 minutes.

And as for the extra weight you've put on, this is a common problem for many teleworkers. You need to exercise (no pun intended) more discipline. Try taking a walk during the time you set aside for lunch. Avoid snacking. Use the time you'd have usually spent commuting to the corporate office in the morning or afternoon instead for a walk, home aerobics video or some other regular physical activity. Not only will this knock off some of the extra weight, but an afternoon or evening workout will help establish an 'end' to the workday and prevent work from creeping into more and more of your personal time. Limits like this are important for teleworkers – because many of the natural demarcation points (such as physically leaving work to go home and getting away from co-workers) are minimized with home offices. And you'll also want to consider the kitchen or food areas 'off limits' for most of the day when you're working at home.

Questions to consider

1. If I were to work full time from home in my current job, what changes would this require in company policy and culture? How would my business need to adjust to support a teleworker?

2. Logistically, what support would I need to work from home? Technologically, what systems or equipment would I need to make this work successfully?

3. Mentally, what adjustments would I need to make as a teleworker? What aspects of this shift in work do I think I'd find easy – and which shifts would be difficult?

Notes

1. WorldCom Conferencing (2001) *Meetings in America III: A Study of the Virtual Workforce in 2001.* The study was commissioned by WorldCom Conferencing and conducted by Modalis Research Technologies. It involved a survey of selected managers and staff from a registered database of volunteers for participation in online research. A copy of the report is available at: http://www.e-meetings.wcom.com/meetingsinamerica/pdf/MIA3.pdf.

2. The report is available at: http://www.computer.org/proceedings/hicss/0001/00015/00015047.pdf.

3. This report draws on the results of the EMERGENCE 18-country employer study and combines them with data from European labour force surveys to develop models of e-working. Four distinct types of 'individual' e-workers are identified in the study: tele-homeworkers, multi-locational e-workers, e-lancers and the e-enabled self-employed. In addition to 'individual' e-working, the report considers the determinants of collective arrangements based on ICT-supported outsourcing or business delocation. The EMERGENCE team has developed forecasts for the spread of each of these individual forms of e-work over the next decade. Based on current patterns, the overall number

of individual e-workers in Europe is likely to triple, from its current levels of over 10 million to 27 million by 2010.

4. From a report prepared for the US Department of Labor in 2000 by the Gartner Group.

5. From a study on telecommuting and employee/employer perceptions by the Boston College Center for Work and Family in 2001.

CHAPTER 8

Facilitating and training virtually

Within virtual organizations, two specific functions have particular implications within a virtual environment: the process of training or employee development and the act of conducting or facilitating meetings. In both cases (training and meeting facilitation), organizations are especially guilty of treating each of these activities as if the virtual nature involves nothing more than the use of some technology to do exactly what had always been done before. Consequently, much of the training that occurs virtually (as either synchronous or asynchronous versions) is terrible and too many staff members can testify to how excruciating a two-hour conference call or videoconference can be.

Understanding facilitation

Let us first start by looking at the rationale and process of facilitation. The role of facilitator tends to vary given the nature of the meeting and the particular challenges the participants present. However, generally speaking we can say that a meeting facilitator is someone who manages the meeting process and interaction so that the participants perform to their best and the meeting achieves its results. Specifically, this means that a facilitator is usually concerned with making sure everyone participates, that some individuals don't do all the talking, that the interaction is consistent with discussion ground rules established by the group, that the conversation stays focused, that action items are captured. A good facilitator also reads non-verbal behaviour for clues to unstated issues and manages disagreements during the meeting.

How is this facilitation role different from that of team leader or manager? There certainly are plenty of cases where the senior person in the room runs a meeting effectively. But typically a facilitator is so focused on process and participant dynamics that they have little time for many of the task and content concerns of a manager. Additionally, position or content authority is frequently a detriment to effective facilitation. An effective facilitator is someone who is perceived as completely objective about the participants and the content (and, accurately or not, most managers are not seen as objective on these issues). Additionally, a good facilitator will generally express no opinion on most if not all of the discussion content – a real challenge for most managers to adhere to. It is this last point that makes it especially problematic for most managers to attempt to facilitate a meeting and do so effectively.[1] Consequently, a good facilitator is often referred to as 'a guide on the side rather than a sage on the stage'[2] because the facilitator may often have very little visible presence within the meeting if many things go right

– intervening only when the meeting threatens to become bogged down or go off course. Typically, a meeting facilitator would try to manage the following issues or dynamics:

- participation – make sure everyone participates and no one dominates conversation;

- focus – make sure the conversation is 'on point' and is neither too deep nor too superficial;

- action items – capture conclusions and 'to dos' and make sure that action items are assigned to individuals or not lost during the meeting;

- conflict and disagreement – use conflict productively and manage interpersonal disputes;

- check for clarity – particularly to verify decisions and watch out for false consensus;

- design processes to achieve group results – creating ice-breakers and decision-methods as necessary or appropriate for given tasks and group dynamics;

- clock-watch – check the time to make sure the group doesn't run over time;

- ground rules – create and/or check meeting norms for the group, enforcing rules as appropriate;

- obstacles – identify barriers to meeting progress, point them out to the group and design processes for avoiding the barriers.

Some facilitators also take on responsibility for meeting set-up and planning (reserving meeting space, coordinating development of the meeting agenda, scheduling meeting participation, room set-up and audio-visual arrangements). While these demands are not rocket science, they do require enough skills and attention that it is difficult to do them well and also participate in the meeting fully. Consequently, for important meetings (especially project kick-offs, initial team get-togethers, strategic planning) it makes sense to bring in a facilitator for the work so that leaders and participants can focus on participating.

Many organizations tend to take the perspective that there is no need to have a designated facilitator for meetings. There are certainly some groups that are organized enough or with meetings that are 'light' enough (in terms of content or challenges to the group) that they succeed without a designated facilitator. However, it is usually true that most meetings benefit from the presence of someone trained in and assigned to manage meeting dynamics. And, it is usually true that most managers or groups fail to understand the value of facilitation – leading to the very common phenomena of inefficient meetings that suck up lots of time and accomplish very little (other than to aggravate the participants).

Facilitators can be internal representatives – a member of the group or team who is designated or volunteers to facilitate the meeting. This has the advantage of being relatively easy logistically (there is no need to contact and coordinate with someone outside the group), the facilitator has background in the group and subject (this is both a positive and negative) and makes it possible (especially if the facilitation role is rotated) for all meetings of longer than 20 minutes to have facilitation. On the complicating side, designating a facilitator isn't enough – the person in that role needs to be clear about what it entails and they need to have the skills to pull it off. Most people aren't good at meeting management. Additionally, being a facilitator means deferring one's opinions on a subject – a frustrating experience for anyone who wants to have 'their say' on a topic. Finally, an internal facilitator may have some 'baggage' with the group. For instance, they may not be perceived as being totally objective or even-handed. Thus for highly volatile subjects, group members or managers (even if trained and experienced in facilitation) may do poorly in this role. An alternative is to engage an external facilitator – or a consultant from outside the organization. This has the advantage (if the choice is right) of bringing in a highly competent and experienced professional who is likely to be (and be perceived) as completely objective about the discussion and the members. However, bringing in outsider facilitators is usually costly (and is therefore difficult to do for all or even most meetings) and also involves a learning curve (where the facilitator is briefed about the topic and meeting participants). As an alternative, a number of organizations have begun to create 'facilitator pools' where groups of employees who volunteer are trained to facilitate and are then available to work with managers on meetings outside of their immediate group. Thus the company has lots of facilitators available at short notice and has little expense to incur in acquiring the facilitator while utilizing someone from within the company but outside the workgroup or team generally provides company knowledge but enough detachment for facilitator objectivity.

Meeting facilitation in virtual settings involves some twists on the previously mentioned skills and requirements. To begin with, as we've discovered in previous chapters, the dynamics of interacting virtually are different from those of face-to-face communication. Thus it's a mistake to assume that what works well face to face will be just as effective in virtual circumstances. Additionally, meeting preparation and set-up become more important in virtual meetings. Meeting processes for virtual discussions often require more structure (because the facilitator or leader may need to make decisions prior to the meeting about which tools to use – such as a nominal group process or an electronic white board) and thus more preparation. Additionally, meetings that involve any kind of technology beg for some degree of contingency planning, i.e. anticipating what to do if something breaks down. Thus the upfront planning requirements for virtual interaction are usually (though not always) greater than face-to-face meetings. Additionally, each meeting tool or technology has some implications for the nature of the meeting, and what tactics and content are likely to be more appropriate given that technology.

Virtual meetings

Like most other virtual work elements, organizations tend to treat virtual meetings as if they were just regular meetings with a little bit of technology. To provide an example, if the sales staff used to regularly meet for two hours every Monday morning, then it shouldn't be a problem for them to shift to a two-hour conference call covering the same subjects, should it? Of course it should.

Virtual meetings have very different dynamics from those of face-to-face meetings. The challenges presented by virtual meetings require participants, managers and facilitators to take different approaches to technology-enabled meetings.

Obviously, conference calls involve situations where participants aren't able to see each other. Yet we tend to assume this shouldn't affect the conversation (or our behaviour). Yet the reverse is true. Normal professionals have trouble focusing for much more than one hour on conference calls. Thus face-to-face meetings on subjects that might run for multiple hours need to be shortened and segmented dramatically if they are converted to audio-conferences. As a rule of thumb, all conference calls should run no longer than one hour. Beyond an hour, almost all people start to have trouble concentrating on a group phone call. The issue here is not the telephone (and the lack of visual stimuli). Two people talking alone on a call are capable of speaking together for more than an hour (a fact you probably know all to well if you have a teenage son or daughter). The issue here is that the nature of a conference call tends to make participants anonymous. Individuals (even if they get an opportunity to talk during the call) tend to feel disembodied. Consequently, conference call participants frequently punch mute buttons on phones (to carry on personal conversations with co-workers next to them while the conference call is going on), step away unannounced to go to the bathroom, or check e-mail in the middle of the call. None of these are behaviours that the same people would consider acceptable in a face-to-face meeting or in a one-on-one call. But the nature of a conference call encourages a perception that their absence won't be missed.

There are several techniques to help make conference calls more productive. As mentioned previously, length matters. Don't let conference calls drag on and on – don't let them run longer than one hour. Another approach is to use a phone bridge.[3] Most business phone systems now have the ability to host conference calls meaning that usually several participants will call in to a particular site (often headquarters). This creates the role of insiders and outsiders. For some reason, this symbolic act (one person hosting the call, the others having to call in) shapes a very powerful subconscious image. People who feel like outsiders in such calls are more likely to tune out during the call (checking e-mail, hitting the mute button or doing other tasks during the call). A phone bridge involves having all participants call into a common number (such as an 800 number) and then punching in a pass code for the meeting. Everyone is an equal. It minimizes such 'tune out' behaviour and enhances a perception of equity. Plus it usually simplifies the process of setting up conference calls. Other than an initial scheduling of the call with the phone bridge provider, everyone simply calls in at the designated time. There is no need to set up the receiving line or for a host to call people individually and put them on hold.

There are also pre-call logistics that make dramatic differences with conference calls. Ideally, all handouts or reference materials for the call should be integrated. If five of us are participating in the call (and each of us submitted material for others to read), that means there are five 'page 1s' (or a separate 'page 1' for each document). Conceivably there are even more 'page 1s' if anyone has sent out multiple documents. Imagine the temporary chaos when Tim says 'on the first page you'll see the projections for the first quarter' and five people sift through a dozen documents looking for the correct page 1. By integrating all of the documents, a reference to 'page 1' or 'page 9' is something everyone can find quickly and easily. Another tip that is very helpful for conference calls (where it is difficult for someone to point to a specific document or indicate with their finger what part of the document they're on) is to use a formatting technique popular with some legal firms. Some legal documents (often transcripts of dispositions) will have each line numbered (a format that most word processing programs support). This format (numbering page lines) is especially effective when conference calls involve long documents. Imagine how easy it would be to follow someone (and how little paper shuffling there would be) if Rita could simply say 'turn to page 12, line 19.' All the call participants would know exactly where she is and what she's referring to.

In a communication exchange that is primarily oral (such as a conference call), it is important to integrate visual elements. In a face-to-face meeting, people can see non-verbal behaviour that indicates if a participant intends to talk or not (such as leaning forward, opening one's mouth, making eye contact, an appearance of anticipation). In a conference call, those non-verbal stimuli are lost. As a consequence, conference calls that involve any discussion or give and take often involve periods of awkward silence, members simultaneously starting to answer and then stopping. This is because the participants can't tell if someone else wishes to speak and are sometimes afraid of interrupting. This can lead to an atmosphere during the call that feels awkward and uneasy. Similarly, if there are a large number of participants, it may be difficult to distinguish voices when people talk. So there are several tips that help deal with these issues. One suggestion is to ask that all participants mention their name when they start to speak during the call. Thus each speaker might say 'this is Tina and I agree that we can accelerate the timeline on this project' or 'Ian here and I think we need more data before we can decide this issue.' This turns disembodied voices into people that participants can 'see' in their minds as they hear the conversation. Another suggestion is visualize a clock and then assign participants to times on the clock.[4] Thus Liam may be '12', Mary is '2', Christina is '4', Malcolm is '6' and so on around the face of the clock. Then the facilitator or team leader would go clockwise to solicit responses or get feedback (you can choose a different starting point each time so the same person isn't always beginning the feedback). This clock approach creates a structure that reduces the awkwardness and uneasiness fostered by participants' uncertainty about who is speaking next. It also allows participants to visualize who will speak next (as well as anticipate their own turn to contribute). Another tactic is for facilitators to help paint visual pictures. For instance, a facilitator might say 'Gordon, I imagine you were shaking your head at that last comment. How can we improve this?' Or, 'Gwen, I can just picture how big your smile is after that report by Dennis. What would you add to his conclusions?' Comments like these paint pictures and build images that turn a conference call into a visual experience as well.

There are a variety of other tips for facilitators of conference calls. Start by taking a roll call. You need to know who is participating – this clarifies who is 'on the call' and participants can also begin to associate names with voices. You may also want to periodically take a roll call later during the call – to see who you may have lost.

It's also important to identify who will take notes. In a face-to-face meeting it's generally obvious if someone is or isn't taking notes, but with a conference call that isn't the case. Participants may assume someone is taking notes but (because behaviour can't be observed) no one is.

The last tip to mention about conference calls is to re-emphasize the limits of a group conference call as a communication means. Conference calls are good tools for sharing information and brief exchanges. But if there are a large number of participants (generally more than five), they break down as means for discussion and debate because it becomes too easy for people to be interrupted or shut out during the conversation. Conference calls work well if they are short, very structured and focus more on information-sharing and/or quick exchanges.

Videoconferencing as a meeting medium

Many people initially regarded videoconferencing as the next best thing to being there. If you couldn't meet face to face, then videoconferencing was supposed to be superior to other alternatives as a meeting forum. This opinion was not based upon actual experience but the assumption that the wider range of sensory options with video – because it wasn't only oral – made it superior to other virtual options. Actual practice has shown this to be quite the opposite. Early videoconferencing technology had significant limits – today's capabilities are much improved. But, despite these improvements, videoconferencing as a virtual meeting technology has some limitations that chat or conference calling do not share.

The initial assumption that the more stimuli the better for a meeting does not seem to apply to videoconferencing as a group interaction technology. Although video does provide visual stimulation, practical experience seems to indicate that other forms of meeting technology perform more effectively. It appears that the combination of picture and voice mediated through technology appears to be distracting. While videoconferencing does provide visual input (which can be particularly helpful for pointing to documents and signposting on visuals), the size of even a large picture on video monitors of a group is still usually too small to help distinguish non-verbal behaviour. While a well-facilitated conference call or chat session seems to make it possible for participants to forget about the presence of the technology and just focus on comments, videoconferencing technology appears to still serve as a distraction from the content of the meeting. For instance, some videoconferencing technology still produces a subtle lag from the point someone speaks to when they're heard by participants (much like analog long-distance phone calls produced decades ago), as if participants were communicating with each other from outer space. Additionally, videoconferencing technology still tends to function best from two points. When there are more than two sites involved, the complexity becomes greater. This may involve a screen with 'picture in picture' in which the independent screens for each location are so small that there might as well be no visual input. Or, if the system is programmed to switch from site to site based upon who speaks first, some locations may get shut out if the home base or default site responds to every comment. I once facilitated a videoconference meeting from five different locations. We eventually had to switch to a conference call because two locations were shut out completely – they would respond to comments but the home site (where the other video feeds were patched into) would have their responses noted first by the hardware and shut out all other comments from other locations until the first site had finished speaking. Approximately 60 per cent of the comments and speaking time were all coming from one site – the home site (where the technology was located) – because the technology would recognize the first site it heard (the local or home site) and cut off sound from the other sites until there was silence. Naturally, this was a frustrating experience as four other sites would respond to a question only to see the video shift to the home site and carry only the home-site comments. In general, for videoconferencing technology to be successful involves much more technical complexity than other alternatives and requires as much (and probably more) structure if the meeting is to function successfully.

However, for those groups that will use videoconferencing for virtual meetings, there are a number of tips to consider, especially for facilitators. Videoconferencing is especially useful as a medium when a virtual team must meet and refer to a number of documents or visuals. While the video picture (especially if it's on a PC or is a picture-in-picture) is usually too small to capture subtle non-verbal signals, it can still be used to help refer participants to the correct document (usually by holding up the book or report you're referring to). This visual element is helpful for many participants.

Videoconferencing requires tremendous upfront planning and preparation. This is generally true of most virtual meeting technologies, but videoconferencing tends to be more planning intensive than most other technologies. Most videoconferencing sites require some advance scheduling (to book time) and it may be necessary to specify the number of sites or hook-ups involved in the meeting.

Participants also need to think about what they wear during the videoconference. Very dark clothing (black or navy for instance) often washes out all detail (so someone in a navy blue suit may look like a head on a dark blob). Depending upon the studio or office background where the videoconference is taking place, the participants may blend into the background (so they appear disembodied) if their clothing is a similar colour to the wall covering or paint. Additionally, video monitors display colours differently depending upon a range of issues (such as software drivers, monitor capability and system settings). Thus a purple pants suit may end up looking like a very garish and disconcerting magenta. Anything with lots of pattern or small detail may end up appearing very 'busy' on the screen as participants see just enough detail to wonder if there is something on the screen or the picture is messed up but not enough to identify what the detail is. Glittery jewellery reflects light that may shine directly into the camera lens (which is both distracting and can create visual problems). Depending upon the distance from the camera, pixel size, picture quality, room lighting and the size of the video screen, people may appear to be people, but it may not be possible to easily distinguish one person from another (especially if the camera must focus on a group of people). Consequently, very large name tents or name tags may be important. Seating participants behind a desk with a modesty panel is a good idea. Otherwise, the placement of the camera may tend to look up women's skirts. Cameras also tend to make people's thighs look significantly larger when they're seated. This may sound like a trivial issue, but remember that when most people hear a tape of their voice, they're thrown by how they sound and it can be very distracting. So when the video creates a distorted picture of the participants (such as 'balloon thighs') that too is often very distracting.

The problem of side conversations becomes greater with videoconferencing. Much like a hearing aid picks up all sound nearby (both the conversation someone is focusing on as well as all the sounds around that the listener isn't focusing on), a videoconferencing sound system picks up all nearby noise. Quick side conversations aren't screened out, although if the camera catches two people whispering but the sound is too quiet to hear, it appears rude and is distracting. On the other hand, if the sound is picked up, it overlaps the primary conversation and produces garble. Additionally, rustling papers, creaking chairs, pens being clicked – these sounds are also often picked up by video sound systems. I can speak from personal experience that you do not want to hear the sound of someone drinking tea or eating a biscuit through the video sound system. So it becomes important for participants to pretend as if they were on a television programme – to watch all actions and sounds, minimize movement, speak clearly or remain silent. Their behaviour needs to be careful and choreographed. If the camera is moving in (or if there is no camera operator and the camera is attached to a PC), then participants need to remain relatively still. In some cases, this means they can't afford to lean forward or backwards, move their seat or stand – because they'll move off camera.

Last of all, expectations are different with videoconferencing. We've grown up in a television age with shorter attention spans.[5]

We're used to being able to switch the channel when we're bored with what we see on the screen. Consequently, by the mere act of adding a video camera, participants will become less patient (and show less attention) than the same players on the same topic would with a conference call or face-to-face meeting. Thus an effective videoconference meeting must move quickly. The facilitator will watch time (and speaker length), cutting people off and enforcing time limits for comments. Hence it is much more important to choreograph a videoconference session than it is an audio conference or chat session. It is useful to view a successful video conference like a television or movie production with all the associated production and preparation details.

Chatting with your fingers

Another form of meeting technology is that of electronic chat. Strictly speaking, chat refers to synchronous text. However, chat has sometimes come to refer to generic group text exchange online. Thus chat is sometimes used to mean asynchronous threaded discussion or asynchronous listserv exchanges as well as synchronous text. Regardless of what form of interaction (synchronous or asynchronous) we're referring to, chat has good potential as a meeting format for particular kinds of discussion.

Synchronous chat has a number of strengths and weaknesses as a meeting technology. Typically, most chat software limits the length of a response an individual can post at any specific moment. The limit is usually a set number of text characters that is the equivalent of a modestly sized paragraph. Some chat programs allow the size of submitted text to be altered (to smaller or larger amounts) though a paragraph is generally the optimal size. This is because if more than three people are participating in the conversation, the conversation and discussion points move quickly enough that by the time someone can type several paragraphs and submit them, the points they're responding to may now be moot – the conversation may have moved onward. Consequently, a paragraph seems to be an optimal size for contributions. Participants who need to send in a response longer than a paragraph typically must send the first paragraph and then immediately send a follow-up response (trying to get the second paragraph submitted before any subsequent responses to avoid interrupting the sequence and flow).

There are a number of tips for synchronous meetings or chat sessions. It helps to pick a regular programme time so people can get in the habit of building it into their schedule. Chat increases the possibility of a greater geographic degree of involvement because you don't need access to a company phone to participate (I've participated in synchronous chat sessions that had participants from Honduras, Singapore, the USA, the UK and Moscow all at the same time). Consequently, it's important to identify time zones (otherwise, when you say 5 p.m., the participant from New York City may think you meant 5 p.m. EST and not 5 p.m. GMT). Additionally, facilitators may want to encourage 'lurkers'[6] to participate and invite people individually to respond.

Synchronous chat works best on limited subjects that don't require detailed responses. Thus it's better to schedule three or four separate chat sessions to cover different topics rather than a longer session to cover all of them in one sitting. It's not that participants have trouble focusing on the chat (like they might having a conference call that runs over an hour) but that so much text gets generated and the nature of the format makes the changes in topics difficult to follow at times. So a focused subject that doesn't require a lot of depth is ideal for synchronous chat. Group votes, checking in or providing quick feedback are also very appropriate for this format.

If the chat room is part of a team or company website, establish an 'ante-room' or 'reading in' centre that participants check into before proceeding to the chat room. This initial entry room can provide any frequently asked questions, some suggestions for operating by chat (if the participants may be new to this form of discussion), meeting ground rules and agenda, or even hyperlinks to background reading. Like face-to-face meetings, some people will show up late. This 'reading in' room provides a chance to bring everyone to the same level of expectations and background regardless of when they join the meeting.

Online meetings that are asynchronous (primarily listserv or threaded discussion) also have a number of strengths and weaknesses. That it is an asynchronous form of discussion makes for a strange meeting dynamic. It works well for very reflective participants or in situations where it is critical to get feedback from select individuals (either to get their buy-in or because only they have the requisite expertise). But because it is asynchronous, it is very easy for participants to procrastinate. Thus some asynchronous meetings may find that most responses occur shortly before the deadline (thus minimizing the ability to read and reflect on the comments).

It is essential for facilitators to carry out extensive anticipation and planning prior to any threaded discussions. A good facilitator will want to anticipate what topics are likely to come up and set up thread topics (and introductions to those topics) prior to the discussion starting. Not only is this important for virtual meetings (the upfront preparation) but it also helps the facilitator keep the conversation focused (the existence of topic areas when participants sign on discourages topics that aren't listed).

There are a number of tips for facilitators of chat sessions (or any virtual meeting). It's important to spend time at the beginning to orient everyone as well as time at the end for wrap-up and summarizing. While this is true even of face-to-face meetings, virtual sessions (like synchronous chat and the asynchronous listserv versions) require more of an effort at tying conversation strands together. When people come online, you need to welcome them and seek to draw them into the conversation (first-time chat participants may have a tendency to 'lurk' rather than jump right in and contribute). Be sure to save the complete transcript of the session – this will be a major plus for knowledge management, generating minutes of the meeting and compiling action items. You can then easily post the transcript elsewhere on the site. Related to this, many organizations and groups are in the habit of inviting outsiders who have particular expertise to participate in team chat sessions. Chat makes this so much easier to do because the invited expert not only doesn't have to be local to participate, they don't even have to be free at an exact time if the meeting is asynchronous. You may want to get in the habit of asking all participants to sign a release allowing you to publish any comments from the chat session – a number of organizations have widely disseminated some of their chat session transcripts and insight. If the transcript is indexed, people can even search it for words or phrases that appear in the transcript.

There are also issues around 'critical mass' when it comes to chat meetings. Some synchronous sessions can be very effective with only a few numbers of participants if the primary purpose is to share information and get some quick input. Otherwise, the basic rule for synchronous and especially asynchronous text-based sessions is that it's very hard to have successful sessions with limited numbers of participants. Synchronous chat can also be very useful as a format for role-plays (such as a mock appraisal between a manager and employee with others 'lurking' in the chat room and then responding after the manager–employee exchange). With critical mass, at some point a listserv or threaded discussion group will literally take off, generating tremendous content and great insight. At such points, these groups develop a real online community with tremendous cohesion and strong social norms.

Training online

There is much that can be written (and has been) about the subject of online learning. Any workforce or organization that involves some degree of virtual interaction is likely to examine (either intentionally or inadvertently) using distance learning within the organization. After all, if some personnel are located 'off-site' then virtual training options need to be developed for those people or they aren't going to be given training and development opportunities by the business. In an age when organizations and governments are increasingly paying more lip service to the importance of intellectual capital and people talent in the role of organizational success, a failure to provide training and development to people is a prescription for competitive failure. Many smaller firms rely on on-the-job training or informal mentoring as a means of development, often assigning a new employee to follow around a veteran to see how it is done. When employees work virtually, such arrangements require much more creativity and typical 'learn by watching others' set-ups tend to fail. Consequently, the priority of developing 'intellectual capital' in a small organization tends to face greater challenges for the typical training approaches that many smaller firms would rely on. Separate from the need to develop personnel, e-learning or distance learning is currently a hot topic. Many organizations have invested millions in their respective currencies to convert traditional classroom training into online formats. Thus for a variety of reasons, organizations that engage in any form of virtual interaction or work need to examine the implications and issues involved with e-learning, distance learning, technology assisted instruction, online training, or virtual staff development (whichever name your organization prefers to use).

As mentioned earlier, the subject of distance learning could form a book in itself. Thus this segment is aimed at providing an introduction to the topic and some guiding principles for any organizations dealing with the subject – especially for how it relates to virtual teaming. Nor will this section of the book provide much advice on technical or hardware matters. The speed with which technology changes and the breadth of this subject area would make it impossible to cover it credibly in any depth. However, there are some key learning and design principles that can be distilled from a range of learning and information efforts online today and applied to efforts tomorrow.

Let us first begin by admitting that most online learning and information today is a vast wasteland. Too many initial online courses consist mostly of text. Books are fine as learning tools for some people. But if a book was sufficient as a teaching tool for most or all learners, there would be no need for training (or even universities). We could simply issue books to all would-be participants and skip the training or courses. Voilà – we'd have a quick solution to how to ensure learning and information dissemination. Sadly, not all (and not even most) adults respond well to learning through text. Additionally, much writing is very poor in nature (think of the last computer manual or documentation you read through cover to cover) so even those predisposed to learn by reading may find that to do so is a dreadful experience given the text they encounter. Finally, some material (regardless of the writing quality or nature of the learner) doesn't lend itself well to writing. This is because what must be acquired is a skill, not necessarily knowledge. For instance, you can read about CPR but that doesn't mean you know how to do it if the person next to you suddenly goes into cardiac arrest. Reading a book on how to land a plane does not mean you're able to accomplish this feat (even if you read the book repeatedly!). Nor does text accurately duplicate the experience or sensation of the event. I've been reliably informed that reading about sex is not quite as satisfactory as actually participating in the act itself. So, despite the love for literature and good writing that many of us share, it (text) is not a satisfactory solution for learning in most cases.

Putting text online further complicates matters. A number of books have been converted to online or e-book formats (to be read on a PDA like a Palm Pilot). While such 'books' appeal to some people (because of accessibility or other convenience issues) they have failed to achieve widespread popularity so far. While some of this may be cultural (such as the joy of a book on one's shelf and the habit of turning pages rather than scrolling), much of the issue involves the challenge of attempting to read text on a video screen. This is exactly the reason why many professionals print out documents from their computer and read the hard copy rather than read from the screen. This is also why companies that attempt to save on paper costs by putting internal documents on websites usually serve only to distribute the paper expenses as almost everyone who accesses the particular document ends up printing out their own hard copy. Basically, people generally find it harder to read text on a computer screen.

The original generation of online courses consisted mostly of online text. Given the previous comments about learning preferences and reading computer screens, it's no wonder that most online learning experiences have been regarded as less than stimulating. A wide range of data from a wide range of sources shows that most online learners fail to complete the course they start. This isn't quite as disappointing as it sounds – online learning (when done well) enables participants to pick and choose the content they need (and ignore the rest), thus resulting in miserable completion rates. After all, why take the last two modules of the course if the stuff you are really interested in is in the first three modules? Still, course completion rates in the single digits also indicate that most online learning participants are less than thrilled with their virtual learning experience.

Virtual learning principles

There are a number of basic operating principles that are key to effective virtual learning, training and employee development efforts. These are true for formal classes (converted to online formats), information sources (such as websites and intranets), FAQ pages (frequently asked question resources), policy and procedure manuals, and any sort of virtual skill or information initiative.

Make it learner centred

This is an incredibly difficult principle for most instructional designers, managers and subject-matter experts to put into practice. I've led plenty of traditional face-to-face classroom training sessions. Let's face it – it's a big ego trip for the instructor. The person at the front of the room is in control. They are the 'expert' who is the centre of attention. They control the agenda, the timing and the sequencing of the course content. Even claims to allow the participants to share in control or to help set the agenda are modest at best. Being a trainer, instructor or presenter in a traditional classroom setting puts you in control. People start when you say it's time to start. The material gets covered at the depth and speed that you see fit. In many cases, you even decide when people take breaks. Not so with online learning. In almost all effective online learning situations, the learner is in control. This means that the learner has the flexibility to decide when to start the course, when to take breaks, what order to take the material, how much time to spend on particular topics, whether to pull in extra (or outside) resources on particular topics, what to repeat or review, and what to skip.

An effective distance course can be a powerful teaching tool because all people are individuals and learn differently. Well-designed distance learning material recognizes this truth and is constructed with it in mind. Thus a well-designed distance course is one in which I should be able to start on module one and jump to module five, you start on the last module and work 'backwards' and yet the course is still extremely effective for us both. That is because we know different material on the same subject (so what you need to learn and what I need to learn is different). We have different learning styles. We face different demands (you must study the material at night because of your current work demands; I must take the course during lunch breaks meaning that I cover small pieces of material at a time). We have different levels of expertise (you took a course in college that had some application to this topic and remember some pieces from that course, I know nothing about the subject). Given these differences, why should we both start at the same starting point, proceed at the same pace, cover all the same topics in the same sequence and be treated as if we were identical twins? We shouldn't and we aren't.

What does learner-centred training look like? Learner-centred material is modular – it is designed so that the learning sequence can vary with each participant (based upon their needs). That means learners need a way to determine what is in a module or chapter and to determine if they really need it or not. Thus good virtual learning will include a clear set of learning outcomes for each module (what the chapter covers and what participants will be able to do after knowing the material). It may include an explanation of why the material is important (or how it fits into a bigger picture). It also includes a means for participants to assess if they know the material (such as a short test before the module). This test can be as simple as (taking an example from online training on quality control) indicating 'you should be able to distinguish between common cause and special cause variation, provide examples of each, and identify three likely sources of common cause variation with plant production lines.'

Appeal to multiple learning styles

People are individuals with individual differences. We don't all learn the same way. There are a variety of different schools of categorization of learning styles. Suffice it to say that regardless of what school you identify with, what works best for some learners doesn't work best for all learners. Effective virtual learning looks for ways to tap into multiple learning styles. Obviously, most online learning can tap into text easily (and thus can appeal very easily to visual or text-based learners). However, with only a little extra work, it is possible to design resources that include mini-video clips or music. Visual applications don't have to be only text – they can include pictures, photos and diagrams. Obviously, you can go overboard on all this, tying up bandwidth and leading to webpages that load very slowly. The idea is not to create glitzy information loaded with clipart, video and other extraneous elements. Instead, you want learning material that doesn't appeal only to readers – it appeals to everyone. Look for ways to design training that works with different learning styles.

Utilize chunking

Many course designers when they hear about the concept of chunking think this means that online learning involves 'dumbing down' the course. This is completely untrue. The concept of chunking means that material needs to be broken down into chunks or pieces of material. A book typically consists of chapters. A training course will consist of modules. Online training needs to be categorized (in part) by screens. As a rule, scrolling is not a good thing. Depending upon graphics, a page that requires scrolling (i.e. does not fit entirely onto one computer screen) will take longer to load (which means it's more likely the learner's attention span will wander). Chunking not only works visually and technically (by making it easier to understand the material and leading to fewer technical problems), it also produces material that is more learner centred. A number of organizations now refer to the concept of 'learning objects' with their online learning. These learning objects consist of training that has been broken up into chunks (or key concepts and skills). The more chunked material is, the easier it is for participants to skip from parts they know to parts they don't. This makes the training more useful (because I'm more likely to return to refresh my memory or test understanding if I know I don't need to repeat the entire course but can go directly to the segment in module six, lesson two, screen three that addresses the skill that I'm worried about).

All online learning must be interactive

Foolishly, many managers and trainers assume that anything that is online (or delivered through technology, such as computer-based training), especially if it is taken individually, is static or one-directional. Good training is interactive – it engages the learner to respond, to provide answers and to think. That is (hopefully) what a trainer or instructor does in a face-to-face classroom when he asks questions of participants or gives them a case study to analyse. Online learning requires interactivity as well. The rule of thumb for the ratio between lessons and interactive elements is 3 to 1. For every three screens of material, the fourth screen should involve an interactive exercise. These interactive elements could involve a short case study (much like the examples at the end of each chapter in this book). They could involve questions for the learner to consider (much like the examples at the end of each chapter in this book). Learners can be asked to 'fill in the blanks' or match alike items. They can be asked to do a search online (usually some kind of a scavenger hunt) in which the first person to answer (or answer with the most correct problems) wins some sort of a prize. Think of the typical employee orientation in which a series of speakers drones on and on and a massive data dump occurs in which the vast majority of the information is not retained by anyone. Imagine instead if the employees were told where they could find the information on particular topics (such as corporate benefits, ethics policy, travel reimbursement procedure and so on), then given a series of questions to answer with the first person answering all ten questions correctly winning a price (perhaps extra vacation time or some item with a company logo on it). Even if the prize is cheap (some sweets), people tend to get engaged with competitions and become interested in doing well. Employees would learn better, orientations would be much shorter and more entertaining, and the organization would have a good idea of what people know and don't know (from the answers the employees turned in for the contest).

What to do?

Imagine that you've just been hired to fill a position because the previous employee has gone overseas. Masha hails from Kazahkstan and has, by all accounts, done a great job for the company. But her family was involved in a serious car accident and she has suddenly left to return to her homeland (and help care for her injured brother and sister). The firm has hired you to fill her spot. Because Masha has already left, you can't observe her work and ask her questions face to face. However, your company has assured you that you'll be able to talk to Masha by phone or through e-mail or other virtual means to get your questions answered. Unfortunately, because Masha worked independently, no one else within the firm can really offer you much advice (other than basic details or generalities) on how she did her job so well. You've looked at possible competitors or other organizations but the option of getting training from someone else doesn't seem very likely. You're going to have to rely on virtual training in this case. How can you get the job training you need in order to succeed if the only alternatives are virtual ones?

The temptation in answering this challenge is to focus solely on the technology and to think about how you might hold conference calls with Masha to get information or set up a video link on laptops so she could observe you at work and provide comments. Those are nice ideas, but the real challenges with this situation have to do with the nature of on-the-job training and its structure, not the technology. If you think about good training and good job direction and then use the technology to get that input from Masha, you'll get the kind of on-the-job training that you need. Some of the information in our earlier chapter about managing dispersed workers is particularly relevant here. You'll want to get guidance from Masha (as well as your management that is located on-site) about the measurable outputs expected from this job. Once you've identified the outputs, these can be broken down into tasks and behaviours that produce the outputs. At first glance, this hypothetical case may have seemed especially challenging. But if you're focusing on outputs, Masha doesn't need to be able to 'see' you to get a sense of how well you're doing and possible areas of breakdown (if your performance is lagging). In addition, you want to first master basic skills or tasks and then build on them. So if one of your tasks involves filling out travel requests online, you might first have Masha show you how to call the form up online and save it before you move to more advanced tasks. You'll also want to have her provide advice in multiple forms. Consequently, you may talk to Masha on the phone, but you'll both want to have your computers up with the same document on the screen and even be holding a hard copy.

Questions to consider

1. Think of the last employee orientation you participated in. What elements of that orientation could have been done virtually? What technology would have been most effective?

2. Think of the best and worst conference calls you've participated in. What made the best conference call so good and what made the worst conference call so bad? What facilitation tips would you identify from these experiences?

3. Imagine that you had to facilitate a virtual meeting using synchronous chat. What facilitation tips do you think would be important for this session to be successful? What work-related topics would lend themselves well to chat and what topics would be better discussed in another format?

Notes

1. There are a number of very good resources on the topic of facilitation in general. On the subject of facilitation and virtual meetings, Lisa Kimball of GroupJazz has several white papers with very specific practical hints at her website (www.groupjazz.com). Nedra Weinstein and Trish Silber with Catalyst Consulting Team have done extensive work with the concept of 'boundaryless facilitation' and I've borrowed from several of their concepts for this chapter. Kimball and Mareen Fisher's book *The Distance Manager* (2001) has some excellent content in their chapters on 'How to use the telephone' (in Chapter 15), 'The distance manager's guide to efficient

teleconferences' (in Chapter 16) and 'Videoconferencing: technology and table manners' (in Chapter 17).

2. George Takacs and Steve Sugar in their book *Games that Teach Teams* (1998) refer to this concept and also provide a very quick and effective explanation of what effective facilitators do.

3. I got this idea from Trish Silber of Catalyst Consulting. Having since used it in a number of meetings of my own, I can vouch for its success.

4. This is a great idea from Lisa Kimball, CEO of GroupJazz, a virtual conferencing and virtual team-building consulting organization. Several white papers and resource guides give further details, available from the website www.groupjazz.com.

5. Lisa Kimball provides more detail on these suggestions for videoconferencing in her white paper *Successful Videoconference Meetings* available at www.groupjazz.com.

6. Someone who 'lurks' in a virtual discussion is an individual who observes but does not post comments. Often a 'lurker' is uncomfortable with the technology, new to the group or slightly intimidated by some other factor. So invitations to participate and post are important. And for longer discussions, allowing new members to initially 'lurk' is sometimes a good policy. They can observe, assimilate group norms, become comfortable, get up to speed with the content and participate when they're ready.

Bibliography

Brent Gallupe et al. (1992) 'Electronic brainstorming and group size', *Academy of Management Journal*, vol. 35, no. 2.

Contu, Diane (1998) 'Trust in virtual teams', *Harvard Business Review*, May–June, pp. 20–1.

Cramton, C. (2002) 'Finding common ground in dispersed collaboration', *Organizational Dynamics*, Spring.

Dangelmaier, W., Forster, D., Horsthemke, V. and Kress, S. (1999) 'Penetration and outlook of telework in Europe: an Internet snapshot, *Proceedings of the 32nd Hawaii International Conference on Systems Sciences.* University of Paderborn, Germany.

DeBono, E. (1999) *Six Thinking Hats* [1986]. Little, Brown.

Duarte, D. and Snyder, N. T. (1999) *Mastering Virtual Teams.* Jossey-Bass.

EMERGENCE Project (2002) *Modelling eWork in Europe: Estimates, Models and Forecasts from the Project*, IES Report 388. Institute of Employment Studies.

Experian (2002) 'Are you an e-mail gossip or a flirt? New survey reveals the online trends of the nation', 9 June. (Survey conducted in conjunction with the Direct Marketing Association (UK)).

Fisher, K. and Fisher, M. D. (2001) *The Distance Manager: A Hands-on Guide to Managing Off-site Employees and Virtual Teams*. McGraw-Hill.

Gibson, Cristina and Cohen, Susan (2003) *Virtual Teams that Work*. Jossey-Bass.

Gilbert, T. (1996) *Human Competence: Engineering Worthy Performance* [1978]. International Society for Performance Improvement.

Grenier, Ray and Metes, Georg (1995) *Going Virtual: Moving Your Organization into the 21st Century*. Prentice-Hall.

Handy, C. (1995) 'How do you manage people whom you do not see?', *Harvard Business Review*, May–June.

Henry, J.E. and Hartzler, M. (1997) *Tools for Virtual Teams*. American Society for Quality.

Hoefling, T. (2001) *Working Virtually*. Stylus.

Jarvenpaa, S. and Leidner, D. (1998) 'Communication and trust in global virtual teams', *Journal of Computer Mediated Communication*, vol. 3, no. 4.

Kimball, Lisa and Digenti, Dori (2001) *Leading Virtual Teams that Learn*. Learning Mastery Press.

Kimball, Lisa and Eunice, Amy (1999) 'The virtual team: strategies to optimise performance', *Health Forum Journal*, May/June, pp. 19–25.

Kroeger, O. (2002) *Type Talk at Work*. Dell Publishing.

Lipnak, J. and Stamps, J. (1997) *Virtual Teams*. John Wiley & Sons.

Mantyla, Karen (1999) *Interactive Distance Learning Exercises that Really Work*. ASTD Press.

Meyerson, D., Weick, K. E. and Kramer, R. M. (1996) 'Swift trust and temporary groups', in R. M. Kramer and T. R. Taylor (eds) *Trust in Organizations: Frontiers of Theory and Research*. Sage.

Naisbitt, J. (1982) *Megatrends*. Warner Books.

Nohria, N. and Eccles, R. G. (1992) *Networks and Organizations: Structure, Form and Action*. Harvard Business School Press.

Ocker, R., Hiltz, S. R., Turoff, M. and Fjermestaad, J. (1996) 'The effects of distributed group support and process structuring on software requirements development teams', *Journal of Management Information Systems*, vol. 12, no. 3, pp. 127–54.

O'Hara-Devereaux, Mary and Johansen, Robert (1994) *Global Work*. Jossey-Bass.

Takacs, G. and Sugar, S (1998) *Games that Teach Teams*. Jossey-Bass.

Salmon, Gilly (2000) *E-Moderating: The Key to Teaching and Learning Online*. Kogan Page.

Thompson, C. (1992) *What a Great Idea*. Harper Perennial.

WorldCom Conferencing (2001) *Meetings in America III: A Study of the Virtual Workforce in 2001*. Research conducted by Modalis Research Technologies.